AF410856

Machine Learning in Image Analysis and Pattern Recognition

Machine Learning in Image Analysis and Pattern Recognition

Editors

Munish Kumar
R. K. Sharma
Ishwar Sethi

MDPI • Basel • Beijing • Wuhan • Barcelona • Belgrade • Manchester • Tokyo • Cluj • Tianjin

Editors

Munish Kumar
Computational Sciences
Maharaja Ranjit Singh Punjab
Technical University, Bathinda
Bathinda
India

R. K. Sharma
Computer Science and
Engineering
Thapar Institute of Engineering
and Technology
Patiala
India

Ishwar Sethi
Computer Science and
Engineering
Oakland University
Oakland
United States

Editorial Office
MDPI
St. Alban-Anlage 66
4052 Basel, Switzerland

This is a reprint of articles from the Special Issue published online in the open access journal *Data* (ISSN 2306-5729) (available at: www.mdpi.com/journal/data/special_issues/Pattern_Recogn).

For citation purposes, cite each article independently as indicated on the article page online and as indicated below:

LastName, A.A.; LastName, B.B.; LastName, C.C. Article Title. *Journal Name* **Year**, *Volume Number*, Page Range.

ISBN 978-3-0365-1714-8 (Hbk)
ISBN 978-3-0365-1713-1 (PDF)

© 2021 by the authors. Articles in this book are Open Access and distributed under the Creative Commons Attribution (CC BY) license, which allows users to download, copy and build upon published articles, as long as the author and publisher are properly credited, which ensures maximum dissemination and a wider impact of our publications.

The book as a whole is distributed by MDPI under the terms and conditions of the Creative Commons license CC BY-NC-ND.

Contents

About the Editors

Munish Kumar

Dr. Munish Kumar received his Master's degree in Computer Science & Engineering from Thapar Institute of Engineering & Technology, Patiala, India in 2008. He received his Ph.D. degree from Thapar Institute of Engineering & Technology, Patiala, India in 2015. He started his career as an Assistant Professor in Computer Science at the Jaito centre of Punjabi University, Patiala. Presently, he is working as an Assistant Professor in the Department of Computational Sciences, Maharaja Ranjit Singh Punjab Technical University, Bathinda, Punjab, India. He has published five international patents. He has published more than 100 research articles in reputed international journals and conference proceedings. He has more than 1000+ citations for his papers on Google Scholar. He is a Professional Member of IEEE. He has guided five Ph.D. research scholars. His research interests include character recognition, handwriting recognition, computer vision, machine learning and pattern recognition.

R. K. Sharma

Professor Rajendra Kumar Sharma received his Ph.D. degree in Mathematics from the University of Roorkee (Now, IIT Roorkee), India, in 1993. He is currently working as a Professor in the Department of Computer Science and Engineering, Thapar Institute of Engineering and Technology, Patiala (India), where he teaches, among other things, statistical models and their usage in computer science. He has been involved in the organization of a number of conferences and other courses at Thapar Institute of Engineering and Technology, Patiala (India). His main research interests are statistical models in computer science, neural networks, and pattern recognition.

Ishwar Sethi

Dr. Ishwar K. Sethi is currently a professor in the Department of Computer Science and Engineering at Oakland University in Rochester, Michigan. He has over 45 years of experience in applying neural networks, including deep learning, machine learning, pattern recognition, and image, video, and text analytics, to a broad range of projects. He has successfully completed projects for the US Air Force, Navy, local hospitals and industries. He has authored or co-authored over 175 journal and conference articles and has graduated 25 doctoral students. He has served on the editorial boards of several prominent journals including IEEE Trans. Pattern Analysis and Machine Intelligence, and IEEE Multimedia. He was elected as an IEEE Fellow in 2001 for his contributions in artificial neural networks and statistical pattern recognition and achieved the status of Life Fellow in 2012.

Preface to "Machine Learning in Image Analysis and Pattern Recognition"

Tens of millions of images are captured every day for a variety of applications in almost all domains of human endeavors. We have come to rely on machine learning, including deep learning, to analyze and extract actionable information from captured images via recognizing the presence of patterns of interest. In recent years, these methods, particularly those using deep learning, have exhibited performance levels surpassing human performance in certain applications. The purpose of this book is to chart the progress in applying machine learning, including deep learning, to a broad range of image analysis and pattern recognition problems and applications. To this end, we have assembled original research articles making unique contributions to the theory, methodology, and applications of machine learning in image analysis and pattern recognition.

Munish Kumar, R. K. Sharma, Ishwar Sethi

Editors

 *data*

Article

An Optimum Tea Fermentation Detection Model Based on Deep Convolutional Neural Networks

Gibson Kimutai [1,*], Alexander Ngenzi [1], Rutabayiro Ngoga Said [1], Ambrose Kiprop [2,3]
and Anna Förster [4]

1   African Center of Excellence in Internet of Things (ACEIoT), College of Science and Technology,
    University of Rwanda, P.O. Box, 3900 Kigali, Rwanda; yngenzi37@gmail.com (A.N.);
    said.rutabayiro.ngoga@gmail.com (R.N.S.)
2   Department of Chemistry and Biochemistry, Moi University, P.O. Box, 3900-30100 Eldoret, Kenya;
    profakiprop@gmail.com
3   African Center of Excellence in Phytochemicals, Textile and Renewable Energy (ACE II-PTRE),
    P.O. Box, 3900-30100 Eldoret, Kenya
4   Sustainable Communication Networks, University of Bremen, 8359 Bremen, Germany;
    anna.foerster@comnets.uni-bremen.de
*   Correspondence: kimutaigibs@gmail.com

Received: 2 April 2020; Accepted: 28 April 2020; Published: 30 April 2020

Abstract: Tea is one of the most popular beverages in the world, and its processing involves a number of steps which includes fermentation. Tea fermentation is the most important step in determining the quality of tea. Currently, optimum fermentation of tea is detected by tasters using any of the following methods: monitoring change in color of tea as fermentation progresses and tasting and smelling the tea as fermentation progresses. These manual methods are not accurate. Consequently, they lead to a compromise in the quality of tea. This study proposes a deep learning model dubbed TeaNet based on Convolution Neural Networks (CNN). The input data to TeaNet are images from the tea Fermentation and Labelme datasets. We compared the performance of TeaNet with other standard machine learning techniques: Random Forest (RF), K-Nearest Neighbor (KNN), Decision Tree (DT), Support Vector Machine (SVM), Linear Discriminant Analysis (LDA), and Naive Bayes (NB). TeaNet was more superior in the classification tasks compared to the other machine learning techniques. However, we will confirm the stability of TeaNet in the classification tasks in our future studies when we deploy it in a tea factory in Kenya. The research also released a tea fermentation dataset that is available for use by the community.

Keywords: machine learning; deep learning; image processing; classification; tea; fermentation

---

1. Introduction

Tea is one of the most popular and lowest cost beverages in the world [1]. Currently, more than 3 billion cups of tea are consumed every day worldwide. This popularity is attributed to its health benefits, which include prevention of breast cancer [2], skin cancer [3], colon cancer [4], neurodegenerative complication [5], prostate cancer [6], and many others. Tea is also attributed to the prevention of diabetes and boosting metabolism [7]. Depending on the manufacturing technique, it may be described as green, black, oolong, white, yellow, and compressed tea [8]. Black tea accounts for approximately 70% of tea produced worldwide. The top four tea-producing countries are China, Sri Lanka, Kenya, and India (Table 1).

Table 1. Top tea-producing countries globally.

Rank	Country	Percentage
1	China	20.6%
2	Sri Lanka	19.3%
3	Kenya	18.2%
4	India	7.5%

Kenya is the largest producer of black tea in the world [7] due to its low altitude, rich loamy soil conditions, ample rainfall, and a unique climate [9]. In Kenya, tea is produced by small- and large-scale farmers. Small-scale farmers are more than 562,000 and account for about 62% of the total tea produced in Kenya [10]. The rest are produced by large-scale tea plantations that operate 39 factories. Smallholder farmers are managed by the government through the Kenya tea development agency (KTDA) board [11]. The board manages 66 tea factories across the country where smallholder tea is processed [1]. Tea is regarded as a significant contributor to the country's economy as it is the leading exchange earner and contributes to more than 4% of the gross domestic product (GDP). The sector is also a source of livelihoods to more than 10% of the country's estimated population of 40 million people [12,13]. Despite the importance of tea to the country, the sector is facing a myriad of challenges which include high production cost, mismanagement, bad agricultural practices, climate change, market competition from other countries, low prices, and lack of automation, among others [13].

There are 5 steps in the production of black tea (Figure 1). The process starts with the plucking of green tea, where two leaves and a bud is the standard. The next step is withering, where tea leaves are spread on a withering bed for them to lose moisture.

Figure 1. Processing steps of black tea.

There is then the cut, tear, and curl step, where tea leaves are cut and torn to open them up for oxidation. The fermentation stage is where tea reacts with oxygen to produce compounds that are responsible for the quality of tea. Heat is passed through tea in the drying stage to remove moisture. The last step is sorting where tea is put into various categories based on their quality. Out of these steps, fermentation is the most important in determining the quality of tea produced [14].

The fermentation process begins when cells of ruptured tea leaves react with oxygen to produce two compounds: Theaflavins (TF) and Thearubins (TR) [15,16]. Theaflavins are responsible for the brightness and briskness of the tea liquor while TR is responsible for the color, taste, and body of tea [16]. During fermentation, the following parameters must be maintained: temperature, relative humidity, and time [15]. The optimum temperature under which fermentation should take place should be approximately 25 °C. The ideal humidity should be approximately 42% [17]. Fermentation is a time-bound process (Figure 2); at the beginning, the liquor is raw and with a green infusion. The formation of TF and TR increases with time until optimum fermentation is achieved. At the optimum fermentation time, the liquor is mellow and with a bright infusion. This is the desired point in fermentation. After optimum fermentation time, the formation of TR reduces and degradation of TF begins. This stage is over-fermentation, where the liquor is soft and with a dark infusion.

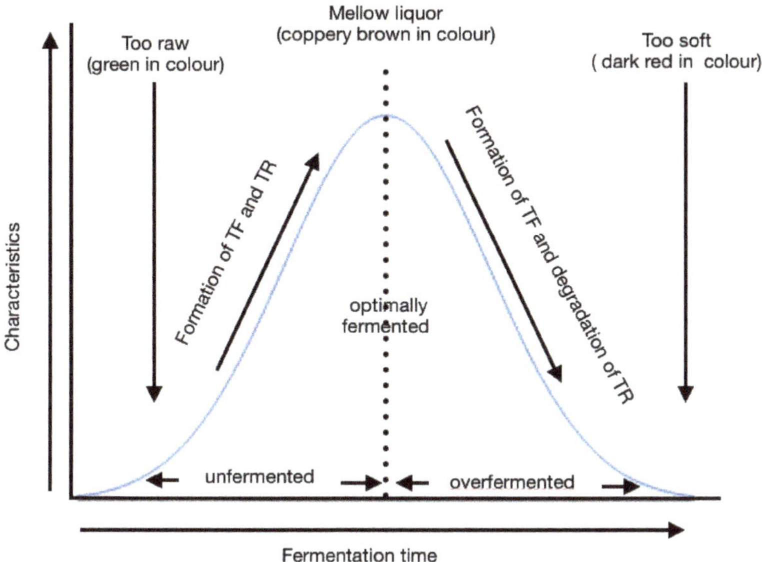

Figure 2. Tea fermentation process.

Currently, tea tasters determine optimum fermentation manually by either of the following methods: smell peaks, color change, infusion, or tasting of tea. The constant intervention of humans in a fermentation room disturbs the environment created for fermentation and is also unhygienic. Moreover, humans are subjective and prone to error [7]. These manual methods lead to a compromise in the quality of produced tea and translate to low prices of tea. Therefore, there is a need for alternative means of monitoring the process of fermentation which is the focus of this research.

Currently, machine learning has been applied to many different fields: engineering, science, education, medicine, business, accounting, finance, marketing, economics, stock market, and law, among others [18–22]. Machine Learning (ML) is a branch of artificial intelligence (AI) that enables a system to learn from concepts and knowledge [23]. Deep learning is a collection of machine learning algorithms which models high-level abstractions in data with nonlinear transformations [24]. Deep learning works with the principle of the Artificial Neural Networks (ANN) system, and its fundamental computation unit is a neuron [19,24,25]. In ML, feature extraction and classification are in different steps, while in deep learning, they are in a single step and are done concurrently.

The contribution of this paper is twofold: First, this research proposes a deep learning model based on CNN for monitoring black tea during fermentation. Secondly, this research releases a tea fermentation dataset [26]. The rest of the paper is arranged as follows: presentation of some of the studies aimed at digitizing fermentation is done in Section 2 and a discussion of materials and methods used in this research is presented in Section 3. We provide implementation of the models and the evaluation metrics in Section 4, while Section 5 provides the results and their discussions. We conclude the study in Section 6.

2. Related Work

With advancements in computing, digitization across many fields is being witnessed [27]. In agriculture, tea processing has been receiving attention from researchers. Proposals have been made on improving the detection of optimum fermentation using the following techniques: electronic nose, electronic tongue, and machine vision. An electronic nose is a smart instrument designed to detect and discriminate odors using sensors [28]. The basic elements of an electronic nose are an odor sensor and an actuator. Proposals to use the electronic nose in detecting optimum tea fermentation have been proposed in References [14,29–32]. In Reference [31], a handheld electronic nose is proposed, while in Reference [32], ultra-low power techniques have been incorporated into an electronic nose. From the literature, it is evident that the electronic nose has made technological breakthroughs. However, they have not been implemented in many tea factories due to its high price and since they are power-hungry. Going into the future, adoption of the electronic nose will depend on innovative ways of using low-cost sensors in their design. It will also depend on the ability to apply ultra-low-power power design techniques to minimize power consumption.

Some studies exist on the application of an electronic tongue to monitor tea fermentation. They include References [33–36]. In Reference [33], an optimal fermentation model with the use of electronic nose and machine learning techniques is proposed. In Reference [34], the authors applied CNN in the development of an electronic tongue. In Reference [37], an electronic tongue to monitor biochemical changes during tea fermentation is proposed. The authors in Reference [35] designed an electronic tongue for testing the quality of fruits. Research in Reference [36] proposes an electronic tongue with the use of a KNN algorithm and adaptive boosting for development. A fusion of the electronic nose and electronic tongue technologies has been proposed in Reference [38]. It is evident from the literature that there have been proposals to use the electronic tongue in detecting optimal fermentation. However, they have not been implemented in tea factories because these technologies are power-hungry and expensive.

The rapid development of computer vision technology in recent years has led to an increased usage of computational image processing and recognition methods. Proposals to apply image processing in the fermentation of tea are reported in References [39–44]. Research in Reference [39] proposes a quality indexing model for black tea during fermentation using image processing techniques. Another remarkable research is in Reference [40], which detects changes in color during fermentation. In Reference [41], artificial neural networks (ANN) and image processing techniques are applied to detect color changes of tea during fermentation. Research in Reference [42] applied SVM with image processing to detect optimum tea fermentation. In Reference [43], the authors used image processing to detect the color change of tea during fermentation. The authors in Reference [44] implemented an electronic tongue with machine vision to predict the optimum fermentation of black tea. From the literature, tea fermentation is an active research area with authors suggesting different approaches. However, the tea fermentation dataset has not to be used. The use of image processing is the most viable approach due to the low cost of imaging devices. Additionally, a color change is easy to detect compared to taste and odor.

3. Materials and Methods

After acquiring data, the next phase was data preprocessing where activities discussed in Section 3.2 were done. The cleaned data was fed to the ML classifiers for training (Figure 3). The training involved hyperparameter tuning until the models were fully trained. Some of the hyperparameters are the learning rate, number of the epoch, regularization coefficient, and batch size. Currently, the available optimization strategies include grid search, random search, hill-climbing, and bayesian optimization, among others [45]. In this study, we adopted the grid search and random search methods. The models were then validated and evaluated using the data discussed in Section 3.1.

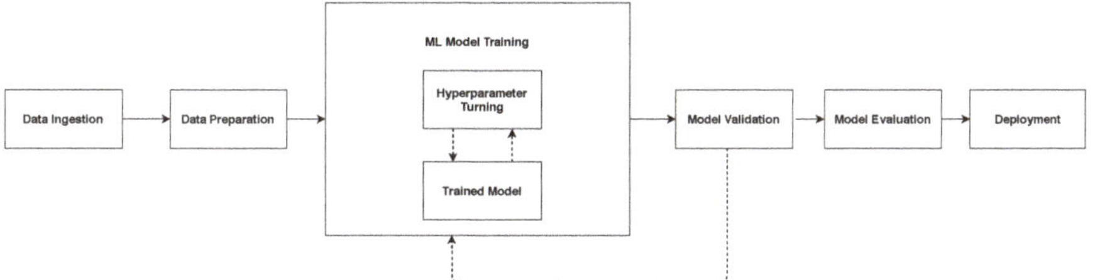

Figure 3. Implementation of machine learning techniques.

In model validation, models which did not pass the validation tests were taken back to the training phase. The evaluation results are presented in Section 5. The models can be deployed to a tea fermentation environment after the aforementioned steps.

This section discusses the datasets used, data preprocessing steps, feature extraction methods, machine learning classification models, and the proposed deep learning model.

3.1. Datasets

In this paper, two datasets were used: tea fermentation and LabelMe datasets. Since there was no existing standard dataset on tea fermentation images existing in the community, we used the LabelMe dataset to validate our results for it is widely used by researchers in image classification to report their results, the dataset is available at no cost, and there was no available dataset on images of tea fermentation images. We discuss each of the datasets in the following paragraphs.

3.1.1. Tea Fermentation Dataset

The images in the tea fermentation dataset [26] were taken in a black tea fermentation environment in a tea factory in Kenya. We used a 5-megapixel camera connected to a Raspberry Pi 3 model B+ to capture the images. Fermentation dataset contains 6000 images that were captured during the fermentation of black tea. Figure 4 shows an image of each of the classes of the tea fermentation dataset. The classes of the images in this dataset are: underfermented, fermented, and overfermented.

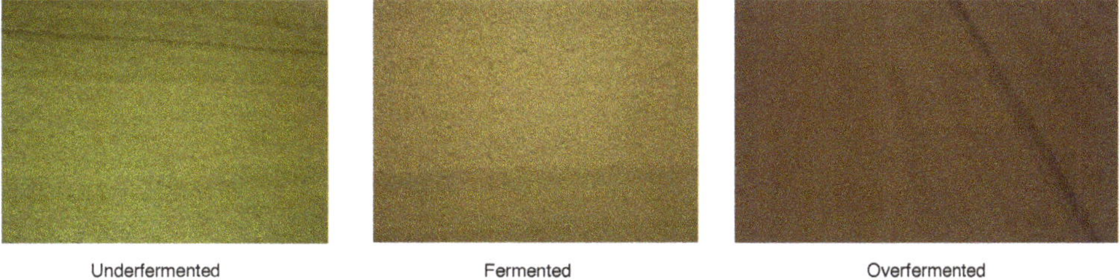

Figure 4. Examples of classes of the tea fermentation dataset.

Table 2 shows the number of images for every class that was used as training, validation, and testing datasets for the classification algorithms. The 80/20 ratio of training/test data is the most commonly used ratio in neural network applications and was adopted in this research. Besides, 10% subset of the test dataset was used to validate the results. A total of 4800 images distributed equally to the 3 classes of images were used for training of the models. To perform validation, 40 images were used in each of the classes while 360 images were used to test the model in each of the 3 classes.

Table 2. The image dataset comprising of three classes of images.

Class	Images Used for Training	Images Used for Validation	Images Used for Testing
Underfermented	1600	40	360
Fermented	1600	40	360
Overfermented	1600	40	360
Total	4800	120	1080

3.1.2. LabelMe Dataset

As explained in Section 3.1, the other dataset that we adopted in this study is the LabelMe dataset [46]. The dataset is one of the standard datasets which researches in the field of image classification use to report their results. The dataset contains 2688 images from 3 classes of outdoor scenes. The classes are forest, coast, and highway. Examples of images from each of the classes are shown in Figure 5.

Figure 5. Examples of categories of LabelMe dataset.

Table 3 shows the number of images used for training, validation, and testing in each of the categories. As with the case in Section 3.1.1, we adopted the 80/20 ratio for training and testing and 10% for validation.

Table 3. Number of images used for training, validation, and testing in the LabelMe dataset.

Class	Images Used for Training	Images Used for Validation	Images Used for Testing
Coast	717	18	161
Forest	717	18	161
Highway	717	18	161
Total	2151	54	483

3.2. Data Preprocessing and Augmentation

After collecting the images as discussed in Section 3.1.1, the images were resized to 150 × 150. Resizing images to 150 × 150 before inputting them into different networks was done to adapt different pretraining CNN structures. We adopted the semantic segmentation annotation method discussed in Reference [47] to annotate the images. There are numerous types of noise in images but the most common are photon noise, readout noise, and dark noise [48,49]. To perform denoising, we adopted the linear filtering method.

3.3. Feature Extraction

Feature extraction in image processing is the process of extracting image features. It is the most crucial step in image classification as it directly affects perfomance of the classifiers [50]. There are various techniques of feature extraction, but in this paper, we adopted color histogram for color feature extraction and Local Binary Patterns (LBP) algorithm for texture extraction.

3.3.1. Color Feature Extraction

Color is an important feature descriptor of an image. During tea fermentation, the color change is evident as the process continues. Relative color histograms in different color spaces can be used to describe tea fermentation images. There are several color spaces which include Red-green Blue (RGB), Hue Saturation Value (HSV), and Hue Saturation Brightness (HSB), among others [51–54]. RGB color space represents a mixture of red, green, and blue. This is the color space that was used to represent the images in this paper. We used color histogram [55] to extract color features of the images that are then fed to the classifiers for training, evaluation, and testing. To construct a feature vector from the color histogram, we used OpenCV [56]. The input was an image of RGB color space. The RGB color space was converted to HSV and represented by 3 channels (the hue, the saturation, and the value). We used 8 bins to represent the three channels. Finally, the range of the channels was between 0–150 since the images had been resized to 150 by 150 pixels. Figure 6a shows an image of underfermented tea, while Figure 6b shows the corresponding color histogram.

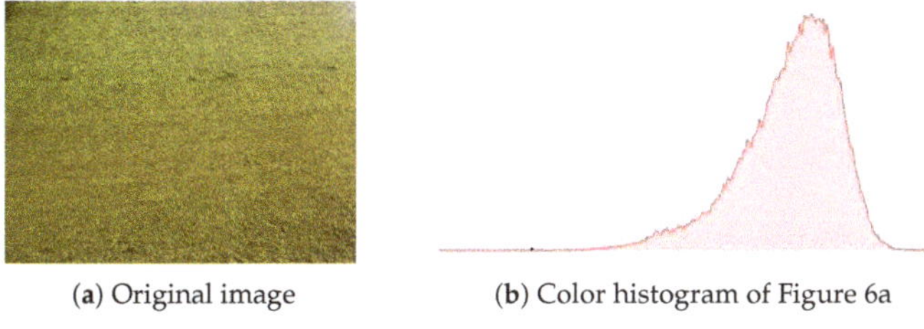

(a) Original image   (b) Color histogram of Figure 6a

Figure 6. Generation of color features of an image using color histogram.

3.3.2. Texture Feature Extraction

Textures are characteristic intensity variations that originate from the roughness of an object surface. The texture of an image is classified into first-order, second-order, and higher-order statistics [57]. There are a variety of methods of extracting texture features including Local Binary Patterns (LBP), the Canny edge detection, discrete wavelet transform, and gray level occurrence matrix, among others [58–60]. In this paper, we adopted LBP to extract the texture features of the images. LBP has many advantages which include reduced histograms and consideration of the center pixels point effect [61], among others. The LBP algorithm is represented by Equation (1):

$$LBP_{x_c,y_c} = \sum_{n=0}^{7} 2^n (I_n - I(x_c, y_c)) \tag{1}$$

where LBP_{x_c,y_c} is the value at the center pixel x_c,y_c, I_n is the values of neighbor pixel, and $I(x_c, y_c)$ is the intensity at the center pixel.

The steps of the texture feature extraction were as follows:

1. The original image was converted into a grayscale image using the approach discussed in Reference [62]. The color grayscale image generated is shown in Figure 7b.
2. LBP algorithm was then used to calculate each of the pixels in the grayscale image as shown in Figure 7. Both LBP_{x_c,y_c} value and texture image are generated. The generated texture image is shown in Figure 7c.
3. Finally, the texture image obtained was converted into gray- scale histogram as shown in Figure 7d.

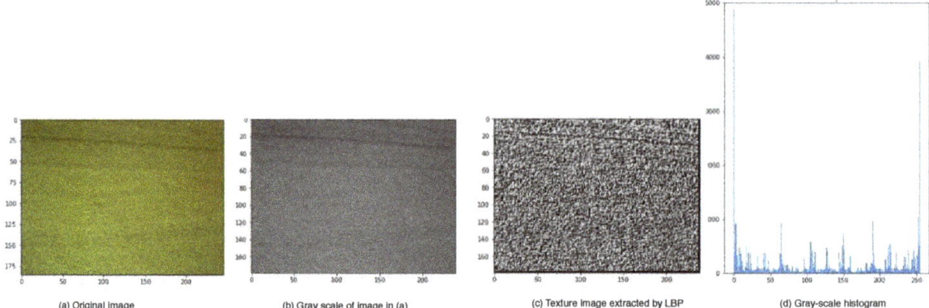

(a) Original image   (b) Gray scale of image in (a)   (c) Texture image extracted by LBP   (d) Gray-scale histogram

Figure 7. Conversion of image to grayscale histogram using Local Binary Patterns (LBP).

3.4. Classification Models

In this paper, we perform the classification of the images in the datasets discussed in Section 3.1 using the following classifiers: decision tree, random forest, K- nearest neighbor, TeaNet, support vector machine, linear discriminant analysis, and naive Bayes. The next paragraphs discuss each of the classifiers.

3.4.1. Decision Tree (DT)

Decision tree is a machine learning technique that employs a tree structure to specify the order of the decisions and the consequences [63]. During training, it generates rules and decision trees. The generated DTs are followed in the classification of the new data [64]. It has the following constituents: root node, internal node, and leaf node (Figure 8). Branches and leaves point to the factors that concern a particular situation [65].

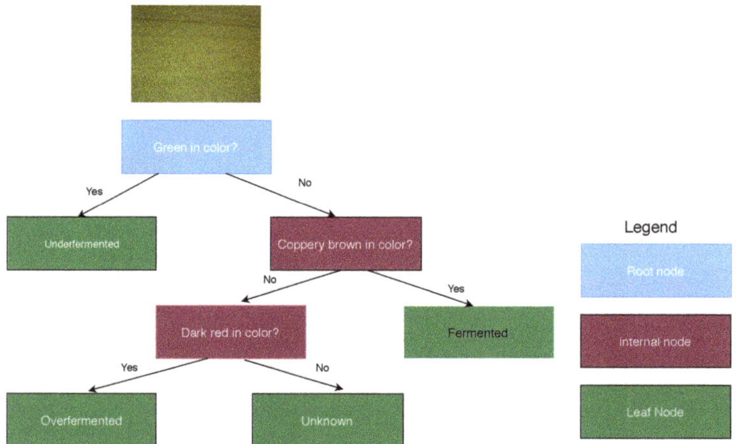

Figure 8. Example of classification by a decision tree.

It is one of the most used machine learning algorithms in classification [66,67] because of its advantages which include high tolerance to multicollinearity [68], flexibility, and exclusion of factors which are not important automatically [63,69], among others. However, DT training is relatively expensive as complexity and time taken are more, a small change in data changes the DT structure, and it is inadequate in predicting continuous values, among others [70].

3.4.2. Random Forest (RF)

Random forest is a machine learning model that operates by constructing multiple decision trees during training [71,72]. The constructed multiple trees are then used for prediction during classification. Each individual tree in the random forest outputs a class prediction and the class with most votes becomes the model's prediction [73]. Figure 9 shows an example of classifying an image using random forest. The image was classified as belonging to class A since the majority of the trees (2) classified the image as belonging to class A. The classifier can estimate missing data, can balance errors in datasets where classes are imbalanced, and can be used for both classification and regression [74–77]. Additionally, it has better results compared to decision tree algorithm; the random forest has a better classification result [72,78]. However, random forest is not as effective in regression tasks as it is in classification and is a black box model [79–81].

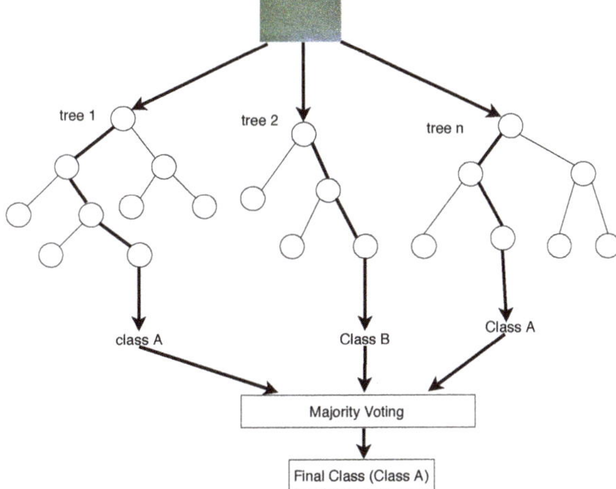

Figure 9. Example of a random forest operation.

3.4.3. K-Nearest Neigbor (KNN)

K-Nearest Neighbor (KNN) is a nonparametric machine learning model used for classification and regression [82–84]. In classification tasks, KNN determines the class of a new sample based on the class of its nearest neighbors. During decision making in a classification task, it finds k training instances that are closest to the unknown instance. It then picks the most occurring classification for the k instances [85]. It determines a dominant category to the target object in which k is the number of training samples. This algorithm assumes that samples close to each other belong to the same category in classification [84]. Figure 10 illustrates an example of a classification using KNN. The task is to find a class that the triangle belongs to. It can either belong to the blue ball class or the green rectangle class. The k is the algorithm we wish to take a vote from. In this case, let us say k = 4. Hence, we will make a cycle with the triangle as the center just to enclose only three data points on the plane. Clearly, the triangle belongs to the blue ball class since all of its nearest neighbors belong to that class.

The algorithm is simple to implement and has a robust search space [86–88]. The main challenge of the model is the expense incurred in terms of large computations in identifying neighbors in a large amount of data [89,90].

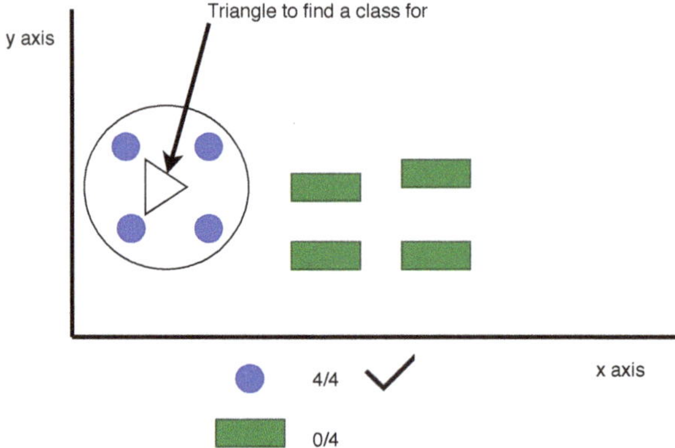

Figure 10. K-Nearest Neighbor (KNN) proximity algorithm map.

3.4.4. Convolutional Neural Network (CNN)

Convolutional Neural Network (CNN) is a class of deep learning technique that is currently emerging in solving computer vision challenges, which includes detection of objects [91], segmentation [92], and dimage classification [93], among others. CNN emerged in the mid-2000s due to the development in computing power of hardware of the computer [94]. A CNN is composed of the following layers (Figure 11): an input layer, convolutional layer, pooling layer, dense layer, and output layer. An input layer of a CNN is the layer where the input is passed to the network. In Figure 11, the input layer contains an image which needs to be classified [95]. Convolutional layers are a set of filters needed to learn. The filters are used to calculate output feature maps, with all units in a feature map sharing the same weights [96–98]. A pooling layer will then sum up the activities and selects the maximum values in the neighborhood of each feature map [99]. A dense layer consists of neurons in a neural network which receive inputs from all the neurons in the previous layer [100]. Convolutional has shown high accuracy in image recognition tasks; however, they have high computation tasks [101].

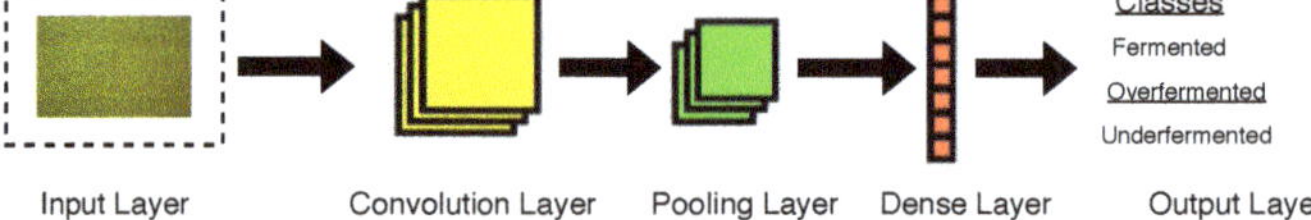

Figure 11. A typical Convolution Neural Network (CNN) architecture.

3.4.5. Support Vector Machine (SVM)

Support Vector Machine (SVM) is a non-probabilistic binary classifier that aims at finding a hyperplane with a maximum margin to separate high dimension classes by focusing on the training samples located at the edge of the class distribution [102]. The model is based on statistical learning theory and the structural risk minimization theory [103]. The model chooses extreme vectors which help in creating the hyperplane. These extreme points are referred to as support vectors. The binary classification problem with linear separability (Figure 12) has a goal to find the optimum hyperplane, through maximizing the margin and through minimizing the classification error between each class.

Some of the advantages of SVM is its ability to rely on its own memory efficiency and its ability to work well with classes having distinct margins [104,105]. However, SVM tends to take a large training time for a large dataset and is not effective for overlapping classes [106,107].

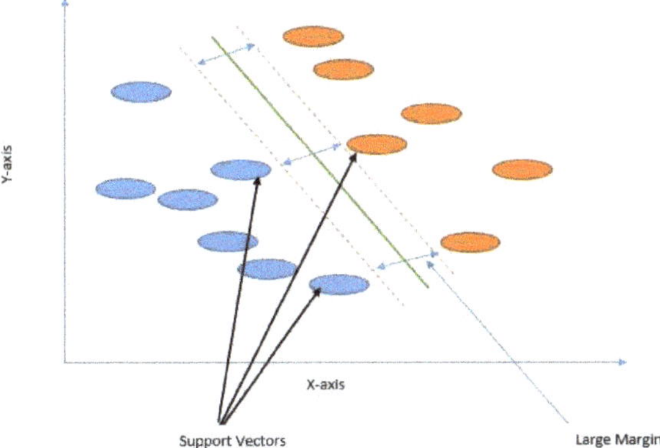

Figure 12. An example of a classification task using Support Vector Machine (SVM).

3.4.6. Naive Bayes (NB)

Naive Bayes is a probabilistic model based on Bayes' theorem. Bayes' theorem provides the relationship between the probabilities of two events and their conditional probabilities [108–110]. A Naive Bayes classifier assumes that the presence or absence of a particular feature of a class is unrelated to the presence or absence of other features [111,112]. In classification tasks, NB constructs a probabilistic model of the features and applies the model in prediction of the new instances. Figure 13 shows a sample of balls belonging to two classes: yellow and green. The task is to estimate the class for which the ball with a question mark belongs to. There is a very high probability that the ball belongs to class green since most of the balls belong to that class.

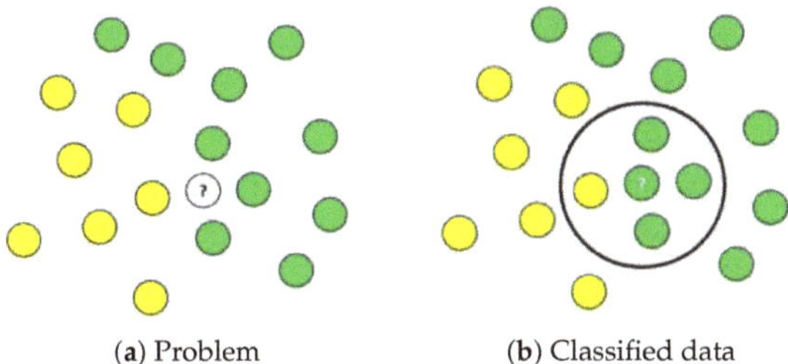

Figure 13. Example of classification using Naive Bayes.

3.4.7. Linear Discriminant Analysis (LDA)

Linear discriminant analysis is an approach developed by the famous statistician R.A. Fisher, who arrived at linear discriminants from a different perspective [113]. He was interested in finding a linear projection for data that maximizes the variance between classes relative to the variance for data from the same class [114]. LDA combines features of a class and builds on separating the classes. It models the differences between classes and builds a vector for differentiating the classes based on the difference in the classes [115,116]. LDA is popular because of its low-cost implementation and its ease of adaptation for discriminating nonlinearly separable classes through the kernel trick method [117], among others. Some of the weaknesses of LDA includes its challenge in handling large datasets, among others [118]. Figure 14a shows a classification problem, while Figure 14b shows the solution to the classification problem using LDA.

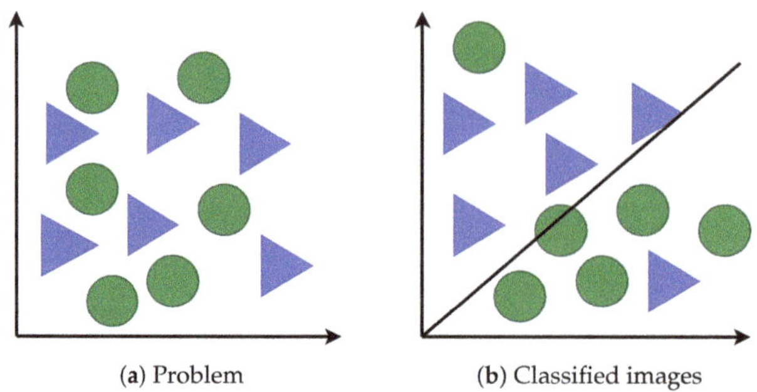

(a) Problem   (b) Classified images

Figure 14. Example of classification using Local Discriminant Analysis (LDA).

3.5. TeaNet

TeaNet is a deep learning model based on Convolutional Neural Network (CNN). The network architecture of TeaNet is an improvement upon the standard AlexNet model [119]. We designed an optimum tea fermentation detection model with relatively simple network structure and small computational needs. To construct TeaNet, we reduced the number of convolutional layer filters and the number of nodes in the fully connected layer. This reduces the number of parameters that require training, thus reducing the overfitting problem. The basic architecture of the network is shown in Figure 15.

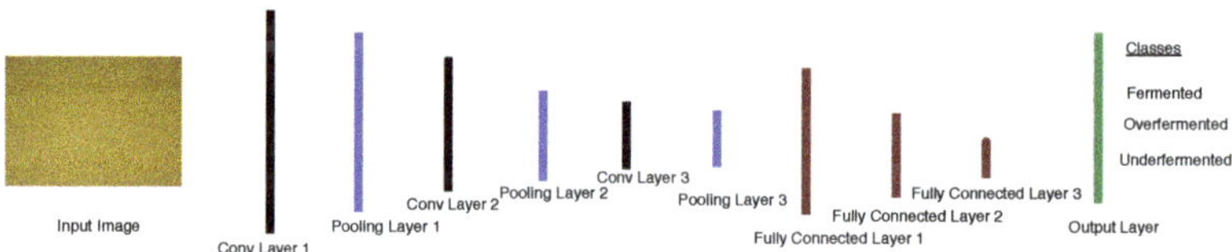

Figure 15. The architecture of the TeaNet that we propose for optimum detection of tea fermentation.

The input images were rescaled to 150×150 pixels, and the three color channels discussed in Section 3.3.1 were all processed directly by the network. Table 4 shows the layer parameters of TeaNet.

Table 4. Layer parameters for TeaNet.

Layer	Parameter	Activation Function
input	$150 \times 150 \times 3$	—
Convolution1 (Conv1)	32 convolution filters (11×11), 4 stride	ReLU
Pooling1 (Pool1)	Max pooling (3×3) 2 stride	—
Convolution2 (Conv2)	64 convolution filters (3×3), 1 stride	ReLU
Pooling2 (Pool2)	Max pooling (2×2) 2 stride	—
Convolution3 (Conv3)	128 convolution filters (3×3), 3 stride	ReLU
Pooling3 (Pool3)	Max pooling (2×2) 2 stride	—
Full Connect4 (fc4)	512 nodes, 1 stride	ReLU
Full Connect5 (fc5)	128 nodes, 1 stride	ReLU
Full Connect5 (fc6)	3 nodes, 1 stride	ReLU
Output	1 node	Softmax

The layers are defined as follows:

1. The first convolutional layer comprises of 32 filters and a kernel size of 11 × 11 pixels. This layer is followed by a rectified linear unit (ReLU) operation. ReLU is an activation function that provides a solution to vanishing gradients [96]. Its pooling layer has a kernel size of 3 ×3 pixels, with two strides.
2. The second convolutional layer comprises of 64 filters and a kernel size of 3 × 3 pixels and is followed by a ReLU operation; its pooling layer has a kernel size of 2 × 2 pixels.
3. Additionally, the third convolutional layer comprises of 128 filters and a kernel size of 3 × 3 pixels, followed by ReLU with a kernel size of 2 × 2 pixels.
4. The first full connection layer was made up of 512 neurons, followed by a ReLu and a dropout operation. The dropout operation [120] is proposed to solve overfitting as it trains only a randomly selected nodes. We set the ratio of dropout to 0.5.
5. The second full convolutional layer had 128 neurons and was followed by a ReLU and dropout operations.
6. The last fully convolutional layer contains three neurons, which represent 3 classes of images in tea fermentation and LabelMe datasets. The output of this layer is transferred to the output layer to determine the class of the input image. A softmax activation function is then implemented to force the sum of the output values to be equal to 1.0. Softmax also limits the individual output values between 0–1.

At the beginning, the weights of the layers were initialized with random values from a Gaussian distribution. To train the network, a stochastic gradient descent (SGD) technique with a batch size of 16 and a momentum value of 0.9 [121] was adopted. Initially, the learning rate across the network was set to 0.1, and a minumum threshold was set at 0.0001. The number of epochs was set as 50, and the weight decay was set to 0.0005. The accuracy of TeaNet increased with an increase in epoch, and it achieved an accuracy of 1.0 at epoch 10 (Figure 16a). At the beginning of the iteration, the accuracy is low since the weights of the neurons are not fully set. After each iteration, the weights are updated. The validation accuracy shows a steady increase, and the model had an accuracy of 1.0. The loss of TeaNet during training and validation is illustrated in Figure 16b. There is a steady reduction in the loss from the first epoch up to epoch 10, where the loss value is at 0 for both the training and validation sets. From Figure 16, the model has good performance and is not overfitted as it records good results in unseen data.

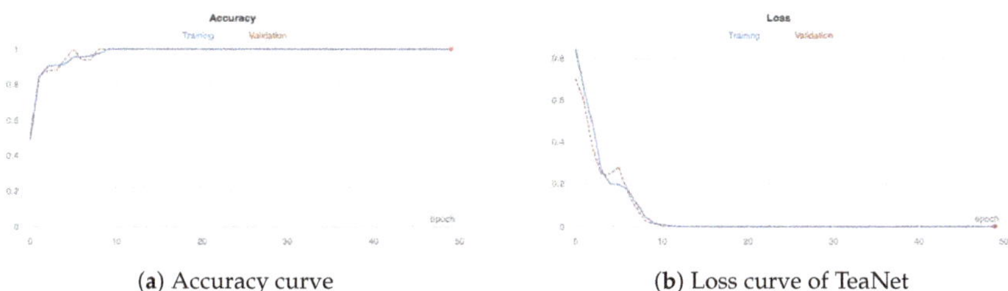

(a) Accuracy curve   (b) Loss curve of TeaNet

Figure 16. Accuracy and loss of TeaNet during training and validation.

4. Implementation

To implement the classification models discussed in Section 3.4 and the TeaNet model discussed in Section 3.5, python programming language was adopted. After implementation, it was necessary to evaluate the performance of the various classification models. This section provides the implementation of the models and the metrics that were adopted in evaluating their performances.

4.1. Implementation of the Classifiers

As mentioned in Section 4, we adopted python programming language to implement the classification models. Some of the reasons for adopting python were that it has rich libraries [122], that it has moderate learning curve [123], and that it is free and open source [124]. Some of the libraries that we adopted alongside python are Tensorflow [125,126], Keras [127], Seaborn [128], matplotlib [129], sklearn [130,131], OpenCV [56], pandas [132], and numpy [133]. The implemented classification models are available at http://classifier.sisiboportal.com/.

4.2. Evaluation Metrics

In Section 4, we mentioned the various evaluation metrics that were adopted in this study to evaluate the classification models. The following paragraphs discuss each of the metric in detail.

4.2.1. Precision

Precision is the ratio of the correct classification to the total number of classifications [134,135]. A low precision indicates a large number of false positives [136]. It can be represented by Equation (2).

$$Precision = \frac{TP}{TP + FP} \tag{2}$$

where TP is an outcome where the model correctly classifies a class and FP is an outcome where the model incorrectly classifies a class.

4.2.2. Recall

Recall is the ratio of the number of correctly classified images to the total number of images [134,136]. It is the actual positives that are correctly classified to the correct classes. Recall can be represented by Equation (3).

$$Recall = \frac{TP}{TP + FN} \tag{3}$$

where TP is an outcome where the model correctly classifies the positive class and FN is an outcome where the model incorrectly classifies the negative class.

4.2.3. F1-Score

F1 Score is the harmonic mean between precision and recall. It tells how precise a classifier is in the classification tasks as well as how robust it is [137]. It is represented by Equation (4).

$$F1 - Score = 2 \times \frac{P \times R}{P + R} \tag{4}$$

where P is the precision and R is the recall.

4.2.4. Accuracy

Accuracy is the fraction of predictions that the model got right. Therefore, it is the sum of correct predictions divided by all the predictions. It can be represented by Equation (5).

$$Accuracy = \frac{TP + TN}{TP + TN + FP + FN} \tag{5}$$

where TP is an outcome where a model correctly classifies the positive class, FP is an outcome where a model incorrectly classifies the positive class, TN is an outcome where the model correctly classifies a negative class, and FN is an outcome where a model incorrectly classifies a negative class.

4.2.5. Logarithmic Loss

Logarithmic loss or log loss works by penalizing false classifications [134]. In classification tasks, it is the measure of the inaccuracy of classification. An ideal logarithmic loss should be 0. Logarithmic loss can be represented by Equation (6).

$$Loss = -(g(log(p)) + (1-g)log(1-p))$$ (6)

where g is the predicted probability and p is the true label.

4.2.6. Confusion Matrix

A confusion matrix is used to summarize the classification performance of a classifier with test data. Sensitivity in a confusion matrix measures the proportion of actual positives that are correctly identified and can be represented by Equation (7).

$$Sensitivity = \frac{TP}{TP + FN}$$ (7)

where TP is the number of correct classification while FN is an outcome where the model incorrectly classifies the negative class.

5. Evaluation Results

In this section, we provide the results of the evaluation of the classification models based on the metrics discussed in Section 4.2.

Results of the precision of the classifiers in the two datasets are shown in Figure 17. All the other classifiers generally categorized the majority of the images correctly. TeaNet classifier clearly performed better than the rest of the classifiers. TeaNet achieved average precisions of 1.0 and 0.96 in the tea fermentation and LabelMe datasets, respectively. Generally, the majority of the classifiers except decision tree produced better precision in the fermentation dataset compared to the LabelMe dataset. This is because there was a distinctive change in color in the 3 categories of fermentation images. The classifiers recorded an average precision of between 0.78–1.00 in the fermentation dataset and between 0.65–0.96 for the LabelMe dataset.

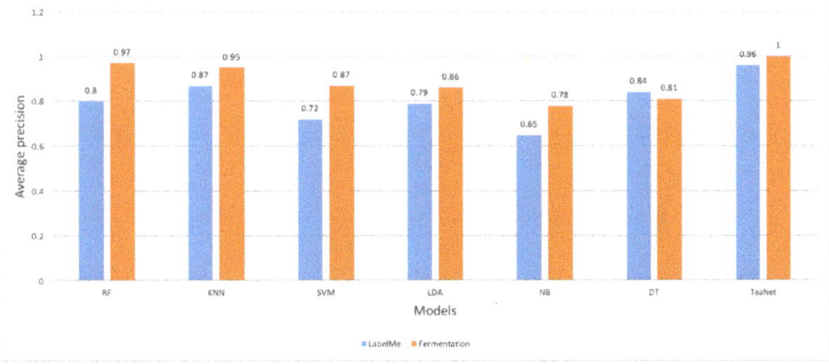

Figure 17. Precision of classification for each of the classifiers for the two datasets.

Recall values are illustrated in Figure 18. Once again, TeaNet outperformed the other classifiers by producing the highest average recall values across the datasets. The majority of the classifiers had better performances in the tea fermentation dataset compared to the LabelMe dataset. The classifiers had an average recall of 0.75–1.0 for the tea fermentation dataset and an average of 0.58–0.96 for LabelMe dataset. KNN also had a good performance by recording average recalls of 0.93 and 0.85 for the tea fermentation and the LabelMe dataset, respectively. Naive Bayes recorded the lowest recall values. From these results, it is evident that TeaNet and KNN produced the best recall values.

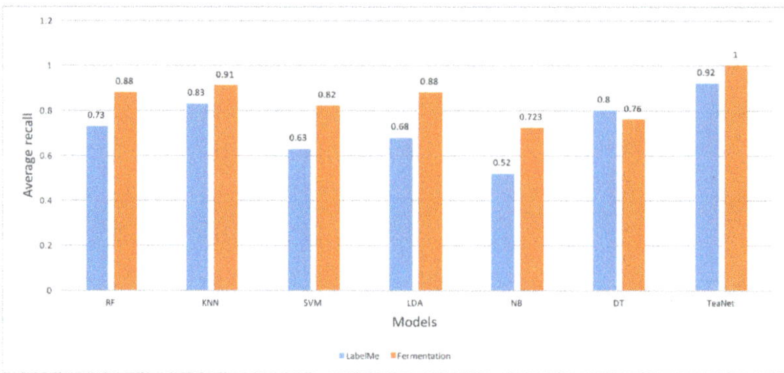

Figure 18. Recall of classification for each of the classifiers for the two datasets.

We compared the F1 score of TeaNet with the other classifiers and presented the results in Figure 19. The F1 score values of TeaNet was higher than the other classifiers. The classifiers recorded F1 values of between 0.58–0.9 for the LabelMe dataset and between 0.75–1.00 for the tea fermentation dataset. We can note that TeaNet showed alot of effectiveness as it achieved an F1 of 1.00 in the tea Fermentation dataset and of 0.9 for the LabelMe dataset (Figure 19). KNN also recorded good performances of 0.93 and 0.85 for tea fermentation and LabelMe datasets, respectively.

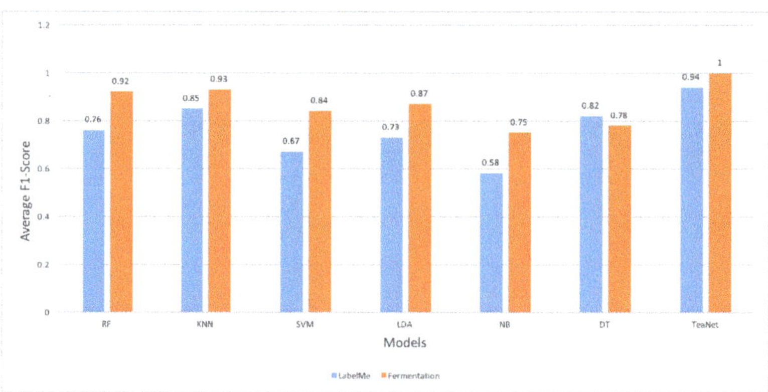

Figure 19. F1 scores of classification for each of the classifiers for the two datasets.

The performance of the classifiers in terms of accuracy is presented in Figure 20. The majority of the classifiers had good accuracy results. TeaNet achieved an average accuracy of 1.00 for the tea fermentation dataset and an average accuracy of 0.958 for the LabelMe dataset. This shows that TeaNet once again outperforms the other classifiers. Each of the classifiers produced an accuracy of more than 0.6 across the datasets. It shows that the probability of each of the classifiers in classifying the dataset is more than 60%. Naive Bayes recorded average accuracies of 0.67 and 0.77 for the LabelMe and tea fermentation dataset, respectively. On the other hand, decision tree recorded average accuracies of 0.94 and 0.85 for the tea fermentation and the LabelMe datasets, respectively. These results show that the majority of the classifiers can be applied to detect the tea fermentation images.

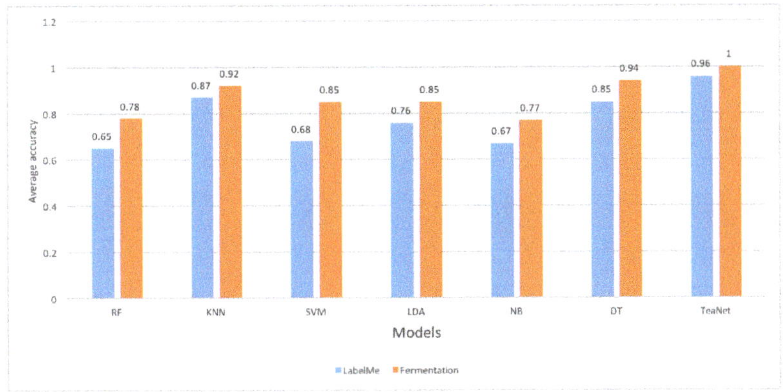

Figure 20. Accuracy of classification for each of the classifiers for the two datasets.

Additionally, logarithmic losses of the classifiers were evaluated. The results of the analysis are shown in Figure 21. TeaNet had the least logarithmic loss at 0.136 and 0.09 for LabelMe and Fermentation dataset, respectively (Figure 21). Generally, the Logarithmic losses recorded by the majority of the models was higher than 0.55 for the LabelMe dataset. For tea fermentation, the majority of the models had logarithmic loss of less than 0.50. Evidently, most of the classifiers had a lower logarithmic loss in the fermentation dataset compared to the LabelMe dataset. Random forest recorded the highest logarithmic loss at 0.7 for the LabelMe dataset. On the other hand, Naive Bayes recorded the highest logarithmic loss of 0.64 for the tea fermentation dataset.

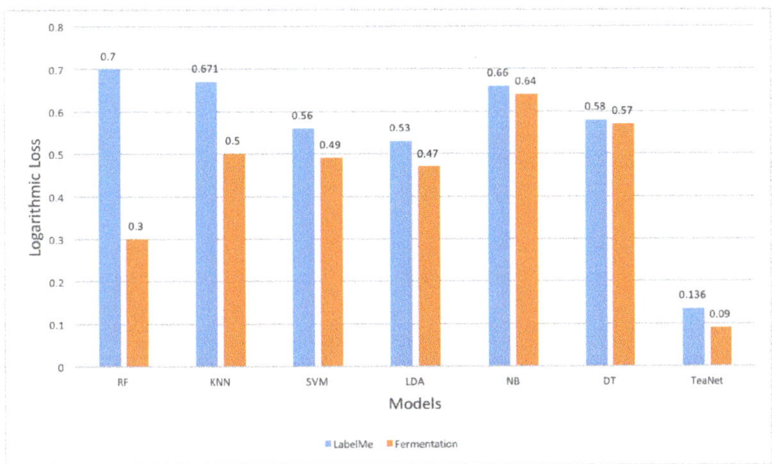

Figure 21. Logarithmic Loss of classification for each of the classifiers for the two datasets.

Finally, a confusion matrix was used to further evaluate the classification models and results are shown in Table 5. The least specificity recorded by the classifiers was 73.5%, and the highest was 100.0%. TeaNet recorded an average sensitivity of 100% for fermented, an average of 100% for overfermented, and finally an average of 100% for underfermented. TeaNet outperformed the other classifiers. Consequently, the TeaNet proposed to classify tea images is superior to the other previously described algorithms.

Table 5. Confusion matrix of the classifiers.

Class	Fermented	Overfermented	Underfermented	Sensitivity
DT (fermented)	250	32	78	69.4%
DT (overfermented)	59	301	0	83.6%
DT (underfermented)	271	0	89	75.3%
SVM (fermented)	296	22	39	82.2%
SVM (overfermented)	68	291	1	80.8%
SVM (underfermented)	61	0	299	83.1%
KNN (fermented)	339	14	7	94.2%
KNN (overfermented)	41	300	19	83.3%
KNN (underfermented)	17	0	343	95.3%
LDA (fermented)	331	11	18	92.0 %
LDA (overfermented)	17	335	8	93.3%
LDA (underfermented)	76	0	284	78.9%
RF (fermented)	325	14	21	90.3%
RF (overfermented)	50	310	0	86.1%
RF (underfermented)	45	0	315	87.5%
NB (fermented)	261	19	80	72.5%
NB (overfermented)	89	253	19	70.3%
NB (under fermented)	96	0	264	73.3%
TeaNet (fermented)	**360**	**0**	**0**	**100.0%**
TeaNet (overfermented)	**0**	**360**	**0**	**100.0%**
TeaNet (underfermented)	**0**	**0**	**360**	**100.0%**

6. Conclusions and Future Work

In this paper, we have proposed a deep learning model dubbed TeaNet. We have assessed the capabilities of TeaNet and other standard machine learning classifiers in categorizing images. We used tea fermentation and LabelMe datasets for training and evaluating the classifiers. From the experimental results, TeaNet outperformed the other classifiers in the classification tasks. In general, all the classifiers had good performance across the two datasets. These results show that the majority of the classifiers can be used in real deployments. Importantly, the effectiveness of TeaNet in the tea fermentation dataset is a great achievement. This is a game changer in the application of deep learning in agriculture and most specifically in tea fermentation.

Additionally, the results from this study highlight the feasibility of applying TeaNet in the detection of tea fermentation, which would significantly improve the process. This will, in turn, increase the quality of produced tea and subsequently increase the value of the made tea. This will lead to improved livelihoods of the farmers and to general improvement of the country's GDP. The same technique can be applied to the fermentation of coffee and cocoa. In our future studies, we will implement TeaNet with the Internet of things in real deployment in a tea factory to monitor fermentation of black tea.

Author Contributions: Formal analysis, A.F.; methodology, G.K. and A.F.; software, G.K.; supervision, A.N., R.N.S., A.K., and A.F.; validation, G.K.; visualization, G.K.; writing—original draft, G.K.; writing—review and editing, A.N., R.N.S., A.K., and A.F. All authors have read and agreed to the published version of the manuscript.

Funding: This research was funded by African Center of Excellence in Internet of Things (ACEIoT) and Kenya Education Network (KENET) through the Computer Science, Information Systems, and Engineering multidisciplinary mini-grant.

Acknowledgments: G.K. wishes to thank A.F. for leading this research, for all the support in ensuring this work is a reality, and for the numerous advises and guidance throughout the research process. G.K. also wishes to put on record appreciation to R.N. and A.N. for their inputs in this work. Additionally, G.K. wishes to express appreciation to A.K. for all the mentorship, support, and guidance. Finally, the authors would like to express their appreciation to the anonymous reviewers and editors for their constructive comments.

Conflicts of Interest: The authors declare that there is no conflict of interest. The funders had no role in the design of the study; in the collection, analyses, or interpretation of data; in the writing of the manuscript; or in the decision to publish the results

Abbreviations

ANN	Artificial Neural Network
CNN	Convolutional Neural Network
SVM	Support Vector Machine
DT	Decision Tree
RF	Random Forests
SVM	Support Vector Machine
KNN	K-Nearest Neighbor
OpenCV	Open Computer Vision
IoT	Internet of Things
TF	Theaflavins
TR	Thearubins
EN	electronic Nose
ET	electronic Tongue
LDA	Local Discriminant Analysis
NB	Naive Bayes
KTDA	Kenya Tea Development Agency
GDP	Gross Domestic Product
MSE	Mean Squarred Error
MAE	Mean Absolute Error

References

1. Sigley, G. Tea and China's rise: Tea, nationalism and culture in the 21st century. *Int. Commun. Chin. Cult.* **2015**, *2*, 319–341. [CrossRef]
2. Yiannakopoulou, E. Green Tea Catechins: Proposed Mechanisms of Action in Breast Cancer Focusing on the Interplay Between Survival and Apoptosis. *Anti-Cancer Agents Med. Chem.* **2014**, *14*, 290–295. [CrossRef]
3. Miura, K.; Hughes, M.C.B.; Arovah, N.I.; Van Der Pols, J.C.; Green, A.C. Black Tea Consumption and Risk of Skin Cancer: An 11-Year Prospective Study. *Nutr. Cancer* **2015**, *67*, 1049–1055. [CrossRef] [PubMed]
4. Hajiaghaalipour, F.; Kanthimathi, M.S.; Sanusi, J.; Rajarajeswaran, J. White tea (Camellia sinensis) inhibits proliferation of the colon cancer cell line, HT-29, activates caspases and protects DNA of normal cells against oxidative damage. *Food Chem.* **2015**, *169*, 401–410. [CrossRef] [PubMed]
5. Sironi, E.; Colombo, L.; Lompo, A.; Messa, M.; Bonanomi, M.; Regonesi, M.E.; Salmona, M.; Airoldi, C. Natural compounds against neurodegenerative diseases: Molecular characterization of the interaction of catechins from green tea with Aβ1-42, PrP106-126, and ataxin-3 oligomers. *Chem. A Eur. J.* **2014**, *20*, 13793–13800. [CrossRef] [PubMed]
6. Gopalakrishna, R.; Fan, T.; Deng, R.; Rayudu, D.; Chen, Z.W.; Tzeng, W.S.; Gundimeda, U. Extracellular matrix components influence prostate tumor cell sensitivity to cancer-preventive agents selenium and green tea polyphenols. In Proceedings of the AACR Annual Meeting 2014, San Diego, CA, USA, 5–9 April 2014; American Association for Cancer Research (AACR): Philadelphia, PA, USA, 2014; p. 232. [CrossRef]
7. Lazaro, J.B.; Ballado, A.; Bautista, F.P.F.; So, J.K.B.; Villegas, J.M.J. Chemometric data analysis for black tea fermentation using principal component analysis. In *AIP Conference Proceedings*; AIP Publishing LLC: Melville, NY, USA, 2018; Volume 2045, p. 020050. [CrossRef]
8. Saikia, D.; Boruah, P.K.; Sarma, U. A Sensor Network to Monitor Process Parameters of Fermentation and Drying in Black Tea Production. *MAPAN* **2015**, *30*, 211–219. [CrossRef]
9. Mitei, Z. Growing sustainable tea on Kenyan smallholder farms. *Int. J. Agric. Sustain.* **2011**, *9*, 59–66. [CrossRef]
10. Kamunya, S.M.; Wachira, F.N.; Pathak, R.S.; Muoki, R.C.; Sharma, R.K. Tea Improvement in Kenya. In *Advanced Topics in Science and Technology in China*; Springer: Berlin/Heidelberg, Germany, 2012; pp. 177–226. [CrossRef]

11. Kagira, E.K.; Kimani, S.W.; Githii, K.S. Sustainable Methods of Addressing Challenges Facing Small Holder Tea Sector in Kenya: A Supply Chain Management Approach. *J. Manag. Sustain.* **2012**, *2*. [CrossRef]
12. Tea Board of Kenya. *Kenya Tea Yearly Report*; Technical Report; Tea Board of Kenya: Nairobi, Kenya, 2018.
13. Onduru, D.D.; De Jager, A.; Hiller, S.; Van Den Bosch, R. *Sustainability of Smallholder Tea Production in Developing Countries: Learning Experiences from Farmer Field Schools in Kenya*; International Journal of Development and Sustainability: Tokyo, Japan, 2012.
14. Xu, M.; Wang, J.; Gu, S. Rapid identification of tea quality by E-nose and computer vision combining with a synergetic data fusion strategy. *J. Food Eng.* **2019**, *241*, 10–17. [CrossRef]
15. Jayabalan, R.; Malbaša, R.V.; Lončar, E.S.; Vitas, J.S.; Sathishkumar, M. A Review on Kombucha Tea-Microbiology, Composition, Fermentation, Beneficial Effects, Toxicity, and Tea Fungus. *Compr. Rev. Food Sci. Food Saf.* **2014**, *13*, 538–550. [CrossRef]
16. Jolvis Pou, K. Fermentation: The Key Step in the Processing of Black Tea. *J. Biosyst. Eng.* **2016**, *41*, 85–92. [CrossRef]
17. Asil, M.H.; Rabiei, B.; Ansari, R. Optimal fermentation time and temperature to improve biochemical composition and sensory characteristics of black tea. *Aust. J. Crop. Sci.* **2012**, *6*, 550–558.
18. Memisevic, R. Deep learning: Architectures, algorithms, applications. In Proceedings of the 2015 IEEE Hot Chips 27 Symposium, HCS 2015, Cupertino, CA, USA, 22–25 August 2015. [CrossRef]
19. Luckow, A.; Kennedy, K.; Ziolkowski, M.; Djerekarov, E.; Cook, M.; Duffy, E.; Schleiss, M.; Vorster, B.; Weill, E.; Kulshrestha, A.; et al. Artificial Intelligence and Deep Learning Applications for Automotive Manufacturing. In Proceedings of the 2018 IEEE International Conference on Big Data (Big Data 2018), Seattle, WA, USA, 10–13 December 2018; pp. 3144–3152. [CrossRef]
20. Shinde, P.P.; Shah, S. A Review of Machine Learning and Deep Learning Applications. In Proceedings of the 2018 4th International Conference on Computing, Communication Control and Automation (ICCUBEA 2018), Pune, India, 16–18 August 2018. [CrossRef]
21. Zhu, W.; Xie, L.; Han, J.; Guo, X. The Application of Deep Learning in Cancer Prognosis Prediction. *Cancers* **2020**, *12*, 603. [CrossRef] [PubMed]
22. Khan, S.; Yairi, T. A review on the application of deep learning in system health management. *Mech. Syst. Signal Process.* **2018**. [CrossRef]
23. Pouyanfar, S.; Sadiq, S.; Yan, Y.; Tian, H.; Tao, Y.; Reyes, M.P.; Shyu, M.L.; Chen, S.C.; Iyengar, S.S. A survey on deep learning: Algorithms, techniques, and applications. *ACM Comput. Surv. (CSUR)* **2018**. [CrossRef]
24. Dargan, S.; Kumar, M.; Ayyagari, M.R.; Kumar, G. A Survey of Deep Learning and Its Applications: A New Paradigm to Machine Learning. *Arch. Comput. Methods Eng.* **2019**, 1–22. [CrossRef]
25. Mahmud, M.; Kaiser, M.S.; Hussain, A.; Vassanelli, S. Applications of Deep Learning and Reinforcement Learning to Biological Data. *IEEE Trans. Neural Netw. Learn. Syst.* **2018**, *29*, 2063–2079. [CrossRef]
26. Gibson, K.; Förster, A. *Black Tea Fermentation Dataset*; Technical Report; Mendeley Ltd.: London, UK, 2020. [CrossRef]
27. Barcelo-Ordinas, J.M.; Chanet, J.P.; Hou, K.M.; García-Vidal, J. *A Survey of Wireless Sensor Technologies Applied to Precision Agriculture*; Wageningen Academic Publishers; Wageningen, The Netherlands, 2013; pp. 801–808.
28. Ghosh, S.; Tudu, B.; Bhattacharyya, N.; Bandyopadhyay, R. A recurrent Elman network in conjunction with an electronic nose for fast prediction of optimum fermentation time of black tea. *Neural Comput. Appl.* **2019**, *31*, 1165–1171. [CrossRef]
29. Bhattacharyya, N.; Seth, S.; Tudu, B.; Tamuly, P.; Jana, A.; Ghosh, D.; Bandyopadhyay, R.; Bhuyan, M. Monitoring of black tea fermentation process using electronic nose. *J. Food Eng.* **2007**, *80*, 1146–1156. [CrossRef]
30. Mohit Sharma, D.G.; Nabarun Bhattacharya. Electronic Nose—A new way for predicting the optimum point of fermentation of Black Tea. *Int. J. Eng. Sci. Invent.* **2012**, *12*, 56–60.
31. Manigandan, N. Handheld Electronic Nose (HEN) for detection of optimum fermentation time during tea manufacture and assessment of tea quality. *Int. J. Adv. Res.* **2019**, *7*, 697–702. [CrossRef]
32. Das, A.; Ghosh, T.K.; Ghosh, A.; Ray, H. An embedded Electronic Nose for identification of aroma index for different tea aroma chemicals. In Proceedings of the International Conference on Sensing Technology (ICST 2012), Kolkata, India, 18–21 December 2012; pp. 577–582. [CrossRef]

33. Ai, S.; Wu, R.; Yan, L.; Wu, Y. Evaluation of taste quality in green tea infusion using electronic tongue combined with LS-SVM. In *Advanced Materials Research*; Trans Tech Publications Ltd.: Stäfa, Switzerland, 2011; Volume 301–303, pp. 1643–1647. [CrossRef]

34. Zhong, Y.h.; Zhang, S.; He, R.; Zhang, J.; Zhou, Z.; Cheng, X.; Huang, G.; Zhang, J. A Convolutional Neural Network Based Auto Features Extraction Method for Tea Classification with Electronic Tongue. *Appl. Sci.* **2019**, *9*, 2518. [CrossRef]

35. Lvova, L. Electronic Tongue Principles and Applications in the Food Industry. In *Electronic Noses and Tongues in Food Science*; Elsevier Inc.: Amsterdam, The Netherlands, 2016; pp. 151–160. [CrossRef]

36. Li, H.; Zhang, B.; Hu, W.; Liu, Y.; Dong, C.; Chen, Q. Monitoring black tea fermentation using a colorimetric sensor array-based artificial olfaction system. *J. Food Process. Preserv.* **2018**, *42*, e13348. [CrossRef]

37. Ghosh, A.; Sharma, P.; Tudu, B.; Sabhapondit, S.; Baruah, B.D.; Tamuly, P.; Bhattacharyya, N.; Bandyopadhyay, R. Detection of Optimum Fermentation Time of Black CTC Tea Using a Voltammetric Electronic Tongue. *IEEE Trans. Instrum. Meas.* **2015**, *64*, 2720–2729. [CrossRef]

38. Zhi, R.; Zhao, L.; Zhang, D. A framework for the multi-level fusion of electronic nose and electronic tongue for tea quality assessment. *Sensors* **2017**, *17*, 1007. [CrossRef]

39. Akuli, A.; Pal, A.; Dey, T.; Bej, G.; Santra, A.; Majumdar, S.; Bhattacharyya, N. Assessment of Black Tea Using Low-Level Image Feature Extraction Technique. In *Proceedings of the Global AI Congress 2019*; Springer: Singapore, 2020; pp. 453–467. [CrossRef]

40. Dong, C.; Liang, G.; Hu, B.; Yuan, H.; Jiang, Y.; Zhu, H.; Qi, J. Prediction of Congou Black Tea Fermentation Quality Indices from Color Features Using Non-Linear Regression Methods. *Sci. Rep.* **2018**, *8*, 10535. [CrossRef]

41. Binh, P.T.; Du, D.H.; Nhung, T.C. Control and Optimize Black Tea Fermentation Using Computer Vision and Optimal Control Algorithm. In *Lecture Notes in Networks and Systems*; Springer: Berlin/Heidelberg, Germany, 2020; Volume 104, pp. 310–319. [CrossRef]

42. Saranka, S.; Thangathurai, K.; Wanniarachchi, C.; Wanniarachchi, W.K. Monitoring Fermentation of Black Tea with Image Processing Techniques. *IPSL* **2016**, *32*, 31–37.

43. Kumar, A.; Singh, H.; Sharma, S.; Kumar, A. Color Analysis of Black Tea Liquor Using Image Processing Techniques. *IJECT* **2011**, *2*, 292–296.

44. Ghosh, A.; Bag, A.K.; Sharma, P.; Tudu, B.; Sabhapondit, S.; Baruah, B.D.; Tamuly, P.; Bhattacharyya, N.; Bandyopadhyay, R. Monitoring the fermentation process and detection of optimum fermentation time of black tea using an electronic tongue. *IEEE Sens. J.* **2015**, *15*, 6255–6262. [CrossRef]

45. Kim, Y.; Chung, M. An Approach to Hyperparameter Optimization for the Objective Function in Machine Learning. *Electronics* **2019**, *8*, 1267. [CrossRef]

46. Li, J.; Wang, J.Z. Automatic linguistic indexing of pictures by a statistical modeling approach. *IEEE Trans. Pattern Anal. Mach. Intell.* **2003**, *25*, 1075–1088. [CrossRef]

47. Tylecek, R.; Fisher, R. Consistent Semantic Annotation of Outdoor Datasets via 2D/3D Label Transfer. *Sensors* **2018**, *18*, 2249. [CrossRef] [PubMed]

48. Fabijańska, A.; Sankowski, D. Image noise removal—The new approach. In Proceedings of the 2007 9th International Conference—The Experience of Designing and Applications of CAD Systems in Microelectronics, Lviv-Polyana, Ukraine, 19–24 February 2007; pp. 457–459. [CrossRef]

49. Ahmad, K.; Khan, J.; Iqbal, M.S.U.D. A comparative study of different denoising techniques in digital image processing. In Proceedings of the 2019 8th International Conference on Modeling Simulation and Applied Optimization (ICMSAO 2019), Zallaq, Bahrain, 15–17 April 2019. [CrossRef]

50. Mistry, Y.; Ingole, D.T.; Ingole, M.D. Content based image retrieval using hybrid features and various distance metric. *J. Electr. Syst. Inf. Technol.* **2017**. [CrossRef]

51. Yang, Y.G.; Zou, L.; Zhou, Y.H.; Shi, W.M. Visually meaningful encryption for color images by using Qi hyper-chaotic system and singular value decomposition in YCbCr color space. *Optik* **2020**, 164422. [CrossRef]

52. Alamgir, N.; Nguyen, K.; Chandran, V.; Boles, W. Combining multi-channel color space with local binary co-occurrence feature descriptors for accurate smoke detection from surveillance videos. *Fire Saf. J.* **2018**, *102*, 1–10. [CrossRef]

53. Li, Z.H.; Han, D.; Yang, C.J.; Zhang, T.Y.; Yu, H.Q. Probing operational conditions of mixing and oxygen deficiency using HSV color space. *J. Environ. Manag.* **2019**, *232*, 985–992. [CrossRef]

54. Garcia-Lamont, F.; Cervantes, J.; López, A.; Rodriguez, L. Segmentation of images by color features: A survey. *Neurocomputing* **2018**, *292*, 1–27. [CrossRef]

55. Novak, C.L.; Shafer, S.A. Anatomy of a color histogram. In Proceedings of the IEEE Computer Society Conference on Computer Vision and Pattern Recognition, Champaign, IL, USA, 15–18 June 1992; Volume 1992-June, pp. 599–605. [CrossRef]

56. Jung, H.; Koo, K.; Yang, H. Measurement-Based Power Optimization Technique for OpenCV on Heterogeneous Multicore Processor. *Symmetry* **2019**, *11*, 1488. [CrossRef]

57. Saroja, G.A.S.; Sulochana, C.H. Texture analysis of non-uniform images using GLCM. In Proceedings of the 2013 IEEE Conference on Information and Communication Technologies (ICT 2013), Thuckalay, India, 11–12 April 2013; pp. 1319–1322. [CrossRef]

58. Kumar, G.; Bhatia, P.K. A detailed review of feature extraction in image processing systems. In Proceedings of the International Conference on Advanced Computing and Communication Technologies (ACCT), Rohtak, India, 8–9 February 2014; pp. 5–12. [CrossRef]

59. Preethi, G.; Sornagopal, V. MRI image classification using GLCM texture features. In Proceedings of the IEEE International Conference on Green Computing, Communication and Electrical Engineering (ICGCCEE 2014), Coimbatore, India, 6–8 March 2014. [CrossRef]

60. Akanbi, O.A.; Amiri, I.S.; Fazeldehkordi, E.; Akanbi, O.A.; Amiri, I.S.; Fazeldehkordi, E. Chapter 4—Feature Extraction. In *A Machine-Learning Approach to Phishing Detection and Defense*; Syngress: Rockland, MA, USA, 2015; pp. 45–54. [CrossRef]

61. Xiaofeng, F.; Wei, W. Centralized binary patterns embedded with image euclidean distance for facial expression recognition. In Proceedings of the 4th International Conference on Natural Computation (ICNC 2008), Jinan, China, 18–20 October 2008; Volume 4, pp. 115–119. [CrossRef]

62. Saravanan, C. Color image to grayscale image conversion. In Proceedings of the 2010 2nd International Conference on Computer Engineering and Applications (ICCEA 2010), Bali Island, Indonesia, 19–21 March 2010; Volume 2, pp. 196–199. [CrossRef]

63. EMC Education Services. *Data Science & Big Data Analytics*; John Wiley & Sons, Inc.: Indianapolis, IN, USA, 2015. [CrossRef]

64. Dahan, H.; Cohen, S.; Rokach, L.; Maimon, O. *Proactive Data Mining with Decision Trees*; SpringerBriefs in Electrical and Computer Engineering; Springer: New York, NY, USA, 2014. [CrossRef]

65. Navada, A.; Ansari, A.N.; Patil, S.; Sonkamble, B.A. Overview of use of decision tree algorithms in machine learning. In Proceedings of the 2011 IEEE Control and System Graduate Research Colloquium (ICSGRC 2011), Shah Alam, Malaysia, 27–28 June 2011; pp. 37–42. [CrossRef]

66. Brunello, A.; Marzano, E.; Montanari, A.; Sciavicco, G. J48SS: A novel decision tree approach for the handling of sequential and time series data. *Computers* **2019**, *8*. [CrossRef]

67. Park, S.; Hamm, S.Y.; Kim, J. Performance Evaluation of the GIS-Based Data-Mining Techniques Decision Tree, Random Forest, and Rotation Forest for Landslide Susceptibility Modeling. *Sustainability* **2019**, *11*, 5659. [CrossRef]

68. Alajali, W.; Zhou, W.; Wen, S.; Wang, Y. Intersection Traffic Prediction Using Decision Tree Models. *Symmetry* **2018**, *10*, 386. [CrossRef]

69. Han, J.; Fang, M.; Ye, S.; Chen, C.; Wan, Q.; Qian, X. Using Decision Tree to Predict Response Rates of Consumer Satisfaction, Attitude, and Loyalty Surveys. *Sustainability* **2019**, *11*, 2306. [CrossRef]

70. Al Hamad, M.; Zeki, A.M. Accuracy vs. cost in decision trees: A survey. In Proceedings of the 2018 International Conference on Innovation and Intelligence for Informatics, Computing, and Technologies (3ICT 2018), Sakhier, Bahrain, 18–20 November 2018. [CrossRef]

71. Feng, D.; Deng, Z.; Wang, T.; Liu, Y.; Xu, L. Identification of disturbance sources based on random forest model. In Proceedings of the 2018 International Conference on Power System Technology (POWERCON 2018), Guangzhou, China, 6–8 November 2018; pp. 3370–3375. [CrossRef]

72. Samarakoon, P.N.; Promayon, E.; Fouard, C. Light Random Regression Forests for automatic multi-organ localization in CT images. In Proceedings of the International Symposium on Biomedical Imaging, Melbourne, Australia, 18–21 April 2017; pp. 371–374. [CrossRef]

73. Georganos, S.; Grippa, T.; Gadiaga, A.; Vanhuysse, S.; Kalogirou, S.; Lennert, M.; Linard, C. An application of geographical random forests for population estimation in Dakar, Senegal using very-high-resolution satellite imagery. In Proceedings of the 2019 Joint Urban Remote Sensing Event (JURSE 2019), Vannes, France, 22–24 May 2019. [CrossRef]

74. Liu, Y.; Liu, L.; Gao, Y.; Yang, L. An improved random forest algorithm based on attribute compatibility. In Proceedings of the 2019 IEEE 3rd Information Technology, Networking, Electronic and Automation Control Conference (ITNEC 2019), Chengdu, China, 15–17 March 2019; pp. 2558–2561. [CrossRef]

75. Saffari, A.; Leistner, C.; Santner, J.; Godec, M.; Bischof, H. On-line random forests. In Proceedings of the IEEE 12th International Conference on Computer Vision Workshops (ICCV Workshops), Kyoto, Japan, 27 September–4 October 2009; pp. 1393–1400. [CrossRef]

76. Kouzani, A.Z.; Nahavandi, S.; Khoshmanesh, K. Face classification by a random forest. In Proceedings of the TENCON 2007—2007 IEEE Region 10 Conference, Taipei, Taiwan, 30 October–2 November 2007. [CrossRef]

77. More, A.S.; Rana, D.P. Review of random forest classification techniques to resolve data imbalance. In Proceedings of the 1st International Conference on Intelligent Systems and Information Management (ICISIM 2017), Aurangabad, India, 5–6 October 2017; Volume 2017-January, pp. 72–78. [CrossRef]

78. Qunzhu, T.; Rui, Z.; Yufei, Y.; Chengyao, Z.; Zhijun, L. Improvement of random forest cascade regression algorithm and its application in fatigue detection. In Proceedings of the 2019 2nd International Conference on Electronics Technology (ICET 2019), Chengdu, China, 10–13 May 2019; pp. 499–503. [CrossRef]

79. Patel, S.V.; Jokhakar, V.N. A random forest based machine learning approach for mild steel defect diagnosis. In Proceedings of the 2016 IEEE International Conference on Computational Intelligence and Computing Research (ICCIC 2016), Chennai, India, 15–17 December 2016. [CrossRef]

80. Xu, B.; Ye, Y.; Nie, L. An improved random forest classifier for image classification. In Proceedings of the IEEE International Conference on Information and Automation (ICIA 2012), Shenyang, China, 6–8 June 2012; pp. 795–800. [CrossRef]

81. Alam, M.S.; Vuong, S.T. Random forest classification for detecting android malware. In Proceedings of the 2013 IEEE International Conference on Green Computing and Communications and IEEE Internet of Things and IEEE Cyber, Physical and Social Computing, Beijing, China, 20–23 August 2013; pp. 663–669. [CrossRef]

82. Thanh Noi, P.; Kappas, M. Comparison of Random Forest, k-Nearest Neighbor, and Support Vector Machine Classifiers for Land Cover Classification Using Sentinel-2 Imagery. *Sensors* **2017**, *18*, 18. [CrossRef]

83. Saadatfar, H.; Khosravi, S.; Joloudari, J.H.; Mosavi, A.; Shamshirband, S. A New K-Nearest Neighbors Classifier for Big Data Based on Efficient Data Pruning. *Mathematics* **2020**, *8*, 286. [CrossRef]

84. Florimbi, G.; Fabelo, H.; Torti, E.; Lazcano, R.; Madroñal, D.; Ortega, S.; Salvador, R.; Leporati, F.; Danese, G.; Báez-Quevedo, A.; et al. Accelerating the K-Nearest neighbors filtering algorithm to optimize the real-time classification of human brain tumor in hyperspectral images. *Sensors* **2018**, *18*, 2314. [CrossRef]

85. Fan, G.F.; Guo, Y.H.; Zheng, J.M.; Hong, W.C. Application of the Weighted K-Nearest Neighbor Algorithm for Short-Term Load Forecasting. *Energies* **2019**, *12*, 916. [CrossRef]

86. Aldayel, M.S. K-Nearest Neighbor classification for glass identification problem. In Proceedings of the International Conference on Computer Systems and Industrial Informatics (ICCSII), Sharjah, UAE, 18–20 December 2012. [CrossRef]

87. Viswanath, P.; Hitendra Sarma, T. An improvement to k-nearest neighbor classifier. In Proceedings of the IEEE Recent Advances in Intelligent Computational Systems (RAICS), Trivandrum, India, 22–24 September 2011; pp. 227–231. [CrossRef]

88. Sun, S.; Huang, R. An adaptive k-nearest neighbor algorithm. In Proceedings of the 2010 7th International Conference on Fuzzy Systems and Knowledge Discovery (FSKD), Yantai, China, 10–12 August 2010; Volume 1, pp. 91–94. [CrossRef]

89. Zhang, S.; Li, X.; Zong, M.; Zhu, X.; Wang, R. Efficient kNN classification with different numbers of nearest neighbors. *IEEE Trans. Neural Netw. Learn. Syst.* **2018**, *29*, 1774–1785. [CrossRef]

90. Jayaraman, P.P.; Zaslavsky, A.; Delsing, J. Intelligent processing of K-nearest neighbors queries using mobile data collectors in a location aware 3D wireless sensor network. In *Lecture Notes in Computer Science (Including Subseries Lecture Notes in Artificial Intelligence and Lecture Notes in Bioinformatics)*; Springer: Berlin/Heidelberg, Germany, 2010; Volume 6098, pp. 260–270. [CrossRef]

91. Cord, A.; Chambon, S. Automatic Road Defect Detection by Textural Pattern Recognition Based on AdaBoost. *Comput.-Aided Civ. Infrastruct. Eng.* **2012**, *27*, 244–259. [CrossRef]

92. Roth, H.R.; Farag, A.; Lu, L.; Turkbey, E.B.; Summers, R.M. Deep convolutional networks for pancreas segmentation in CT imaging. *Med. Imaging 2015 Image Process.* **2015**, *9413*, 94131G. [CrossRef]

93. Wang, W.; Wu, B.; Yang, S.; Wang, Z. Road Damage Detection and Classification with Faster R-CNN. In Proceedings of the 2018 IEEE International Conference on Big Data (Big Data 2018), Seattle, WA, USA, 10–13 December 2018; pp. 5220–5223. [CrossRef]

94. Radopoulou, S.C.; Brilakis, I. Automated Detection of Multiple Pavement Defects. *J. Comput. Civ. Eng.* **2017**, *31*. [CrossRef]

95. Chen, Q.; Fu, Y.; Song, W.; Cheng, K.; Lu, Z.; Zhang, C.; Li, L. An Efficient Streaming Accelerator for Low Bit-Width Convolutional Neural Networks. *Electronics* **2019**, *8*, 371. [CrossRef]

96. Véstias, M.P. A Survey of Convolutional Neural Networks on Edge with Reconfigurable Computing. *Algorithms* **2019**, *12*, 154. [CrossRef]

97. Chen, C.; Ma, Y.; Ren, G. A Convolutional Neural Network with Fletcher–Reeves Algorithm for Hyperspectral Image Classification. *Remote Sens.* **2019**, *11*, 1325. [CrossRef]

98. Song, J.; Gao, S.; Zhu, Y.; Ma, C. A survey of remote sensing image classification based on CNNs. *Big Earth Data* **2019**, *3*, 232–254. [CrossRef]

99. Cire\csan, D.C.; Meier, U.; Masci, J.; Gambardella, L.M.; Schmidhuber, J. Flexible, High Performance Convolutional Neural Networks for Image Classification. In Proceedings of the Twenty-Second International Joint Conference on Artificial Intelligence (IJCAI'11), Barcelona, Catalonia, Spain, 16–22 July 2011; Volume Two, pp. 1237–1242.

100. Paoletti, M.; Haut, J.; Plaza, J.; Plaza, A. Deep&Dense Convolutional Neural Network for Hyperspectral Image Classification. *Remote Sens.* **2018**, *10*, 1454. [CrossRef]

101. Hinton, G.; Deng, L.; Yu, D.; Dahl, G.; Mohamed, A.R.; Jaitly, N.; Senior, A.; Vanhoucke, V.; Nguyen, P.; Sainath, T.; et al. Deep neural networks for acoustic modeling in speech recognition: The shared views of four research groups. *IEEE Signal Process. Mag.* **2012**, *29*, 82–97. [CrossRef]

102. Elaidi, H.; Elhaddar, Y.; Benabbou, Z.; Abbar, H. An idea of a clustering algorithm using support vector machines based on binary decision tree. In Proceedings of the 2018 International Conference on Intelligent Systems and Computer Vision (ISCV 2018), Fez, Morocco, 2–4 April 2018; Volume 2018-May, pp. 1–5. [CrossRef]

103. Chao, C.F.; Horng, M.H. The construction of support vector machine classifier using the firefly algorithm. *Comput. Intell. Neurosci.* **2015**, *2015*. [CrossRef]

104. Cheng, G.; Tong, X. Fuzzy clustering multiple kernel support vector machine. In Proceedings of the International Conference on Wavelet Analysis and Pattern Recognition, Chengdu, China, 15–18 July 2018; Volume 2018-July, pp. 7–12. [CrossRef]

105. Shahbudin, S.; Zamri, M.; Kassim, M.; Abdullah, S.A.C.; Suliman, S.I. Weed classification using one class support vector machine. In Proceedings of the 2017 International Conference on Electrical, Electronics and System Engineering (ICEESE 2017), Kanazawa, Japan, 9–10 November 2017; Volume 2018-January, pp. 7–10. [CrossRef]

106. Kranjčić, N.; Medak, D.; Župan, R.; Rezo, M. Support Vector Machine Accuracy Assessment for Extracting Green Urban Areas in Towns. *Remote Sens.* **2019**, *11*, 655. [CrossRef]

107. Wang, D.; Peng, J.; Yu, Q.; Chen, Y.; Yu, H. Support Vector Machine Algorithm for Automatically Identifying Depositional Microfacies Using Well Logs. *Sustainability* **2019**, *11*, 1919. [CrossRef]

108. Karthika, S.; Sairam, N. A Naïve Bayesian classifier for educational qualification. *Indian J. Sci. Technol.* **2015**, *8*. [CrossRef]

109. Walia, H.; Rana, A.; Kansal, V. A Naïve Bayes Approach for working on Gurmukhi Word Sense Disambiguation. In Proceedings of the 2017 6th International Conference on Reliability, Infocom Technologies and Optimization: Trends and Future Directions (ICRITO 2017), Noida, India, 20–22 September 2017; Volume 2018-January, pp. 432–435. [CrossRef]

110. Jahromi, A.H.; Taheri, M. A non-parametric mixture of Gaussian naive Bayes classifiers based on local independent features. In Proceedings of the 19th CSI International Symposium on Artificial Intelligence and Signal Processing (AISP 2017), Shiraz, Iran, 25–27 October 2017; Volume 2018-January, pp. 209–212. [CrossRef]

111. Harahap, F.; Harahap, A.Y.N.; Ekadiansyah, E.; Sari, R.N.; Adawiyah, R.; Harahap, C.B. Implementation of Naïve Bayes Classification Method for Predicting Purchase. In Proceedings of the 2018 6th International Conference on Cyber and IT Service Management (CITSM 2018), Parapat, Indonesia, 7–9 August 2018. [CrossRef]

112. Dewi, Y.N.; Riana, D.; Mantoro, T. Improving Naïve Bayes performance in single image pap smear using weighted principal component analysis (WPCA). In Proceedings of the 3rd International Conference on Computing, Engineering, and Design (ICCED 2017), Kuala Lumpur, Malaysia, 23–25 November 2017; Volume 2018-March, pp. 1–5. [CrossRef]

113. Zhao, H.; Wang, Z.; Nie, F. A New Formulation of Linear Discriminant Analysis for Robust Dimensionality Reduction. *IEEE Trans. Knowl. Data Eng.* **2019**, *31*, 629–640. [CrossRef]

114. Singh, A.; Prakash, B.S.; Chandrasekaran, K. A comparison of linear discriminant analysis and ridge classifier on Twitter data. In Proceedings of the IEEE International Conference on Computing, Communication and Automation (ICCCA 2016), Noida, India, 29–30 April 2016; pp. 133–138. [CrossRef]

115. Jerkovic, V.M.; Kojic, V.; Popovic, M.B. Linear discriminant analysis: Classification of on-surface and in-air handwriting. In Proceedings of the 2015 23rd Telecommunications Forum (TELFOR 2015), Belgrade, Serbia, 24–26 November 2015; pp. 460–463. [CrossRef]

116. Ghosh, J.; Shuvo, S.B. Improving Classification Model's Performance Using Linear Discriminant Analysis on Linear Data. In Proceedings of the 2019 10th International Conference on Computing, Communication and Networking Technologies (ICCCNT 2019), Kanpur, India, 6–8 July 2019. [CrossRef]

117. Shashoa, N.A.A.; Salem, N.A.; Jleta, I.N.; Abusaeeda, O. Classification depend on linear discriminant analysis using desired outputs. In Proceedings of the 2016 17th International Conference on Sciences and Techniques of Automatic Control and Computer Engineering (STA 2016), Sousse, Tunisia, 19–21 December 2016; pp. 328–332. [CrossRef]

118. Markopoulos, P.P. Linear Discriminant Analysis with few training data. In Proceedings of the ICASSP, IEEE International Conference on Acoustics, Speech and Signal Processing, New Orleans, LA, USA, 5–9 March 2017; pp. 4626–4630. [CrossRef]

119. Sun, J.; Cai, X.; Sun, F.; Zhang, J. Scene image classification method based on Alex-Net model. In Proceedings of the 2016 3rd International Conference on Informative and Cybernetics for Computational Social Systems (ICCSS 2016), Jinzhou, China, 26–29 August 2016; pp. 363–367. [CrossRef]

120. Kim, P. *MATLAB Deep Learning: With Machine Learning, Neural Networks and Artificial Intelligence*; Apress: New York, NY, USA, 2017; p. 162. [CrossRef]

121. Sutskever, I.; Martens, J.; Dahl, G.; Hinton, G. On the Importance of Initialization and Momentum in Deep Learning. In Proceedings of the 30th International Conference on International Conference on Machine Learning (ICML'13), Atlanta, GA, USA, 16–21 June 2013; Volume 28, pp. III–1139–III–1147.

122. Dubosson, F.; Bromuri, S.; Schumacher, M. A python framework for exhaustive machine learning algorithms and features evaluations. In Proceedings of the International Conference on Advanced Information Networking and Applications (AINA), Crans-Montana, Switzerland, 23–25 March 2016; Volume 2016, pp. 987–993. [CrossRef]

123. Samal, B.R.; Behera, A.K.; Panda, M. Performance analysis of supervised machine learning techniques for sentiment analysis. In Proceedings of the 2017 3rd IEEE International Conference on Sensing, Signal Processing and Security (ICSSS 2017), Chennai, India, 4–5 May 2017; pp. 128–133. [CrossRef]

124. Tohid, R.; Wagle, B.; Shirzad, S.; Diehl, P.; Serio, A.; Kheirkhahan, A.; Amini, P.; Williams, K.; Isaacs, K.; Huck, K.; et al. Asynchronous execution of python code on task-based runtime systems. In Proceedings of the ESPM2 2018: 4th International Workshop on Extreme Scale Programming Models and Middleware, Held in conjunction with SC 2018: The International Conference for High Performance Computing, Networking, Storage and Analysis, Dallas, TX, USA, 12 November 2018; pp. 37–45. [CrossRef]

125. Stancin, I.; Jovic, A. An overview and comparison of free Python libraries for data mining and big data analysis. In Proceedings of the 2019 42nd International Convention on Information and Communication Technology, Electronics and Microelectronics (MIPRO 2019), Opatija, Croatia, 20–24 May 2019; pp. 977–982. [CrossRef]

126. Abadi, M.; Barham, P.; Chen, J.; Chen, Z.; Davis, A.; Dean, J.; Devin, M.; Ghemawat, S.; Irving, G.; Isard, M.; et al. TensorFlow: A system for large-scale machine learning. *arXiv* **2016**, arXiv:1605.08695.

127. Fahad, S.K.; Yahya, A.E. Big Data Visualization: Allotting by R and Python with GUI Tools. In Proceedings of the 2018 International Conference on Smart Computing and Electronic Enterprise (ICSCEE 2018), Shah Alam, Malaysia, 11–12 July 2018. [CrossRef]

128. Baker, B.; Pan, L. Overview of the Model and Observation Evaluation Toolkit (MONET) Version 1.0 for Evaluating Atmospheric Transport Models. *Atmosphere* **2017**, *8*, 210. [CrossRef]

129. Hung, H.C.; Liu, I.F.; Liang, C.T.; Su, Y.S. Applying Educational Data Mining to Explore Students' Learning Patterns in the Flipped Learning Approach for Coding Education. *Symmetry* **2020**, *12*, 213. [CrossRef]

130. Pedregosa, F.; Varoquaux, G.; Gramfort, A.; Michel, V.; Thirion, B.; Grisel, O.; Blondel, M.; Prettenhofer, P.; Weiss, R.; Dubourg, V.; et al. Scikit-learn: Machine learning in Python. *J. Mach. Learn. Res.* **2011**, *12*, 2825–2830.

131. Zeng, Z.; Gong, Q.; Zhang, J. CNN model design of gesture recognition based on tensorflow framework. In Proceedings of the 2019 IEEE 3rd Information Technology, Networking, Electronic and Automation Control Conference (ITNEC 2019), Chengdu, China, 15–17 March 2019; pp. 1062–1067. [CrossRef]

132. Li, G.; Liu, Z.; Cai, L.; Yan, J. Standing-Posture Recognition in Human–Robot Collaboration Based on Deep Learning and the Dempster–Shafer Evidence Theory. *Sensors* **2020**, *20*, 1158. [CrossRef]

133. Momm, H.G.; ElKadiri, R.; Porter, W. Crop-Type Classification for Long-Term Modeling: An Integrated Remote Sensing and Machine Learning Approach. *Remote Sens.* **2020**, *12*, 449. [CrossRef]

134. Kosmopoulos, A.; Partalas, I.; Gaussier, E.; Paliouras, G.; Androutsopoulos, I. Evaluation Measures for Hierarchical Classification: A unified view and novel approaches. *Data Min. Knowl. Discov.* **2013**. [CrossRef]

135. Flach, P. Performance Evaluation in Machine Learning: The Good, the Bad, the Ugly, and the Way Forward. *Proc. AAAI Conf. Artif. Intell.* **2019**, *33*, 9808–9814. [CrossRef]

136. Tharwat, A. Classification assessment methods. *Appl. Comput. Inform.* **2018**. [CrossRef]

137. Hai, M.; Zhang, Y.; Zhang, Y. A Performance Evaluation of Classification Algorithms for Big Data. *Procedia Comput. Sci.* **2017**, *122*, 1100–1107. [CrossRef]

© 2020 by the authors. Licensee MDPI, Basel, Switzerland. This article is an open access article distributed under the terms and conditions of the Creative Commons Attribution (CC BY) license (http://creativecommons.org/licenses/by/4.0/).

Data Descriptor

LeLePhid: An Image Dataset for Aphid Detection and Infestation Severity on Lemon Leaves

Jorge Parraga-Alava [1],*, Roberth Alcivar-Cevallos [1], Jéssica Morales Carrillo [2], Magdalena Castro [1], Shabely Avellán [1], Aaron Loor [2] and Fernando Mendoza [2]

[1] Facultad de Ciencias Informáticas, Universidad Técnica de Manabí, Avenida Jose María Urbina, Portoviejo 130104, Ecuador; roberth.alcivar@utm.edu.ec (R.A.-C.); Magdalena.castro@utm.edu.ec (M.C.); Shabely.avellan@utm.edu.ec (S.A.)

[2] Carrera de Computación, Escuela Superior Politécnica Agropecuaria de Manabí, Sitio El Limón, Calceta 130250, Ecuador; jmorales@espam.edu.ec (J.M.C.); aaron.loor@espam.ec (A.L.); fernando.mendoza@espam.ec (F.M.)

* Correspondence: jorge.parraga@usach.cl; Tel.: +593-9-6731-7778

Abstract: Aphids are small insects that feed on plant sap, and they belong to a superfamily called *Aphoidea*. They are among the major pests causing damage to citrus crops in most parts of the world. Precise and automatic identification of aphids is needed to understand citrus pest dynamics and management. This article presents a dataset that contains 665 healthy and unhealthy lemon leaf images. The latter are leaves with the presence of aphids, and visible white spots characterize them. Moreover, each image includes a set of annotations that identify the leaf, its health state, and the infestation severity according to the percentage of the affected area on it. Images were collected manually in real-world conditions in a lemon plant field in Junín, Manabí, Ecuador, during the winter, by using a smartphone camera. The dataset is called LeLePhid: lemon (Le) leaf (Le) image dataset for aphid (Phid) detection and infestation severity. The data can facilitate evaluating models for image segmentation, detection, and classification problems related to plant disease recognition.

Dataset: https://doi.org/10.17632/tndhs2zng4

Dataset License: CC-BY

Keywords: plant disease recognition; image segmentation; image classification; aphid; *Aphoidea*; lemon

check for **updates**

Citation: Parraga-Alava, J.; Alcivar-Cevallos, R.; Morales Carrillo, J.; Castro, M.; Avellán, S.; Loor, A.; Mendoza, F. LeLePhid: An Images Dataset for Aphids Detection and Infestation Severity on Lemons Leaf. *Data* **2021**, *6*, 51. https://doi.org/10.3390/data6050051

Academic Editors: Munish Kumar, R. K. Sharma and Ishwar Sethi

Received: 14 April 2021
Accepted: 12 May 2021
Published: 17 May 2021

Publisher's Note: MDPI stays neutral with regard to jurisdictional claims in published maps and institutional affiliations.

Copyright: © 2021 by the authors. Licensee MDPI, Basel, Switzerland. This article is an open access article distributed under the terms and conditions of the Creative Commons Attribution (CC BY) license (https://creativecommons.org/licenses/by/4.0/).

1. Summary

The dataset, called LeLePhid in short, provides images of lemon leaves. This dataset contains 665 photos of the top and back of lemon tree leaves in which there are healthy and unhealthy leaves; these were collected manually in citrus crops from Junín, Ecuador, in winter, from December to May, when the weather is warm and rainy in this country. For the annotation process, it was carried out with the Labelbox© annotation tool, and to assign the severity of the infestation, three annotators manually inspected the image and set the grade of infestation severity according to [1] and the OIRSA method [2]. These data can be used for training, testing, and validation of computational models related to image segmentation and object detection in plant disease studies. At the same time, they can be helpful for researchers and professionals working on computer vision-based models for image classification and object detection using images of healthy leaves and leaves with the presence of aphids. The data annotations can be used to develop and improve the accuracy of lemon leaf aphid infestation severity and detection algorithms.

2. Data Description

The LeLePhid dataset provides lemon leaf images that can be used to develop and evaluate the performance of models of image segmentation, object detection, and classification problems related to plant diseases. The dataset contains imagery of the upper and back sides of leaves of lemon trees manually collected in citrus crops around Junín, Ecuador. On each image, the foreground leaf is identified, and its status is labeled, i.e., healthy and aphid[1] presence. The dataset also includes annotations to identify the infestation severity of the leaves affected by aphids. It can be used to design automatic aphid counting models because, as stated in [3], compared with manual counting[2], these models can calculate the percentage of the affected area through analyzing the image information. The released files for the so-named LeLePhid dataset are two folders: the raw data are available in the "Images" folder (665 images of lemon leaves) and pre-processed data are available in the "Annotation" folder (.json and .xlsx files). Samples of them are depicted in Figures 1 and 2. Figure 1 shows an example of the annotated images for segmentation purpose. In a green limited-area is identified the a lemon leaf. In purple areas the aphids presence. In Figure 2A, the class of the image is healthy meanwhile in Figure 2B the class is aphids, i.e., the leaf has presence of this insect. In addition, Tables 1 and 2 describe the levels or infestation severity on each lemon leaf available in the dataset. Finally, Figure 3 describes the distribution of images by health status and levels of infestation of aphids.

Figure 1. Annotation examples of a segmentation mask in the LeLePhid dataset.

---

[1] Aphids are tiny insects that feed by sucking sap from plants, and they can cause diseases.
[2] An adhesive board is placed on the plants, and researchers count the aphids on it.

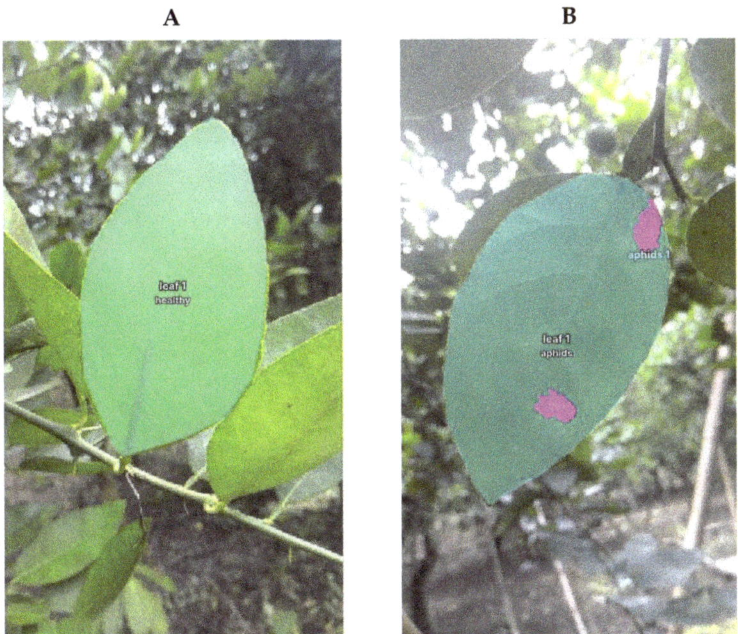

Figure 2. Annotation examples of lemon leaf image classification.

Distribution of leaf infestation severity levels

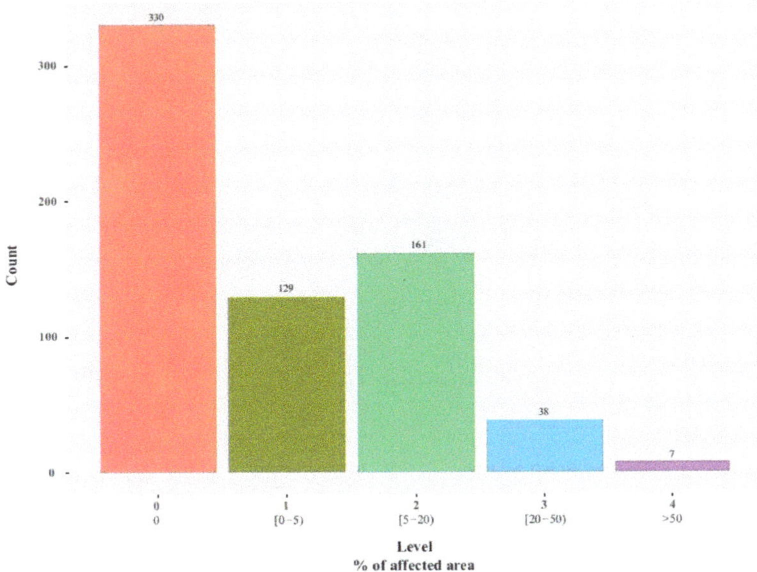

Figure 3. Distribution of images according to infestation severity levels.

3. Methods

LeLePhiD is designed to support computer vision research related to image processing, with a particular focus on the detection and infestation severity of aphids on lemon leaves. The pipeline of the creation of this dataset is shown in Figure 4.

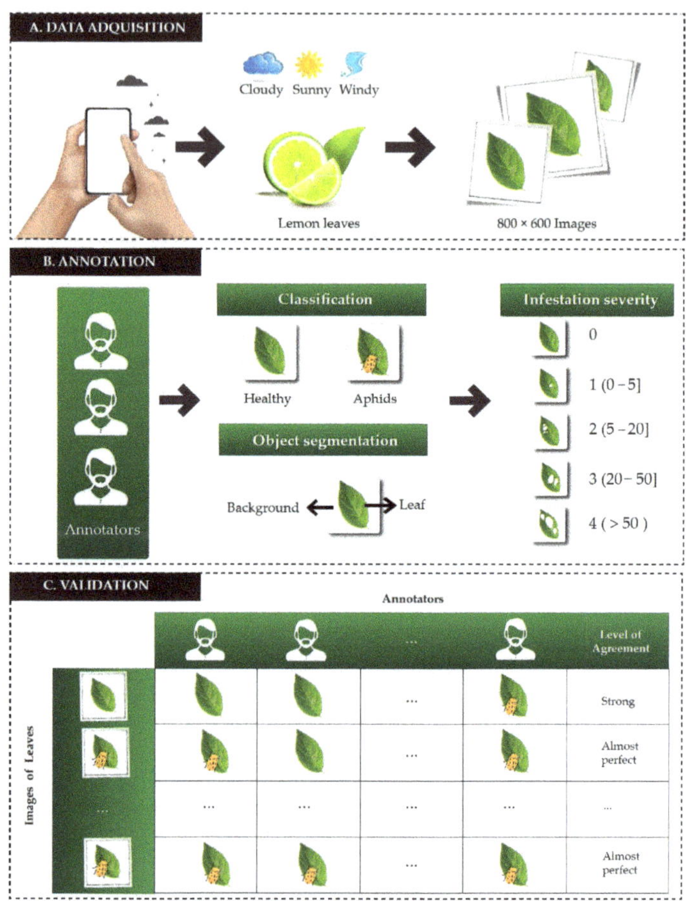

Figure 4. Pipeline of creation of LeLePhid dataset.

In Figure 4, we can see three steps, including the data acquisition, the incorporation of annotations, and the validation. In the following subsections, we detail each part of the process of creating this dataset.

3.1. Data Acquisition

The lemon leaf images were manually acquired on a crop field in a rural area of Junín, Manabí, by using a 2 megapixel smartphone camera. Lemon images were captured following the procedure in [4] during cloudy, sunny, rainy, and windy days. The images were taken at a distance of 30–50 cm from the plant. The data capture process was performed in a time window of two weeks with different climatic conditions and background scenarios. We took 665 leaf images of the upper and back sides of healthy and unhealthy lemon plants. All images were rotated to a vertical position and resized to 800 × 600 pixels, keeping the aspect ratio. The process can be observed in Figure 4A.

3.2. Annotations

The annotation process was performed by using the Labelbox© annotation tool and can be observed in Figure 4B. In the object segmentation annotation, for each image, the foreground leaf is identified, and also, if the leaf is diseased, the area affected by aphids is marked (Figure 1). In the classification annotation, each image is labeled healthy or aphids according to the leaf health status (Figure 2). These annotations were assigned based on the comprehensive evaluation of the images of leaves according to the experience of the annotators.

Figure 1 shows an example of an annotation where the green-limited area identifies a lemon leaf and the magenta areas show the presence of aphids.

In Figure 2, the labeled lemon leaf images are shown. In Figure 2A, the image class is healthy; meanwhile, in Figure 2B, the class is aphids, i.e., the leaf has the presence of this insect. Note that only certain areas with white spots and texture correspond to aphids. Other spots related to other leaf conditions are not considered in this study.

To assign the infestation severity of each leaf, three annotators manually inspected the image and set the grade of infestation severity according to [1] and the OIRSA method [2]. The description of the levels of infestation severity grades of the affected area in lemon leaves can be observed in Table 1.

Table 1. Infestation severity scale of aphids in plants.

Level	% Affected Area	Symptom
0	0	Healthy plant with no aphid presence.
1	[0–5)	Few aphids. Foliage with no yellowing symptoms.
2	[5–20)	Crinkling and curling of few leaves of the plant.
3	[20–50)	Crinkling and curling of leaves almost all over the plant.
4	>50	Extreme curling, crinkling, and drying all over the plant.

3.3. Validation

The validation process can be observed in Figure 4C. The consistency of the annotations was validated using the agreement between annotators. This was achieved by seeking matches in the category assigned to each image by the annotators. To quantify this, we used the kappa coefficient and the interpretation suggested by [5]. It can be simplified in Table 2 as follows:

Table 2. Interpretation of Cohen's kappa.

Kappa	Level of Agreement	% of Data Reliability
0–0.20	None	0–4%
0.21–0.39	Minimal	4–15%
0.40–0.59	Weak	15–35%
0.60–0.79	Moderate	35–63%
0.80–0.90	Strong	64–81%
Above 0.90	Almost Perfect	82–100%

In Table 2, any kappa value below 0.60 indicates inadequate agreement among the annotators and that little confidence should be placed in the labeling process. Here, the percentage of data reliability corresponds to the squared kappa value. The final value of each label (level) was selected using a plurality strategy, i.e., when the matches are greater than 2. In cases of ties, the value was arbitrarily chosen in random order. The level of agreement obtained by our annotators was 91.0%, which means that the real percentage of affected area by aphids with LeLePhid is almost perfect.

Finally, the LeLePhid dataset contains 665 lemon leaf images distributed into 330 healthy leaves and 335 leaves with aphid presence. The latter are categorized according to leaf infestation severity and distributed as summarized in Figure 3.

In Figure 3, we include the image distribution according to the infestation severity levels. Note that there are 330 images with 0% of affected area, i.e., they correspond to healthy photos. The 335 images with aphid presence are divided into four levels. The first one has 129 leaf images with less than 5% of affected area (level 1). Most of the leaf images (161) are of level 2, i.e., they have between 5% and 20% of affected area. Further, there are 38 images with between 20% and 50% of infestation (level 3). Finally, the dataset contains seven leaf images with more than 50% of affected area (level 4).

4. User Notes

The data described in this paper are from a citrus crop near Junín, Ecuador (latitude −0.9277, longitude −80.2058). They were acquired using a smartphone camera. The identifications of the leaf, its state, and the area affected by aphids were individually incorporated as annotations over the image. The annotation is provided as a JSON file supported in any computer vision software. The possibilities of practical application are the following:

- The data can be used to train, test, and validate computational models related to image classification on plant disease studies. In this sense, we already have evidence from a previous work [6], where convolutional neural networks (CNNs) were used to board a binary classification problem related to lemon leaves with aphid presence. The quality of LeLePhid was evidenced by allowing the model to achieve average rates between 81% and 97% of correct aphid classification.
- The data can be helpful to researchers and professionals working on computer vision-based models for image segmentation and object detection using images of healthy leaves and leaves with aphid presence. Cases such as those discussed in [3,7] are examples of the potential that our dataset can offer from the point of view of continuous improvement of machine learning algorithms to address segmentation and identification problems related to plant diseases.
- The data can serve as a motivation to encourage further research into the agriculture sector and computer vision methods for citrus pest identification. Image annotation is the data labeling technique used to make the varied objects recognizable for computers. Our dataset includes image annotations of leaves and aphid-infected areas to make them recognizable or even understandable for computers. These annotations can be used to help the large-scale monitoring of the health of crops through, for instance, devices such as UAVs (unmanned aerial vehicles) or drones, where works in [8–11] have already demonstrated the benefits that can be obtained in the agricultural sector when devices such as drones are used in conjunction with computer vision.

Note that most of the images used by algorithms of the two first bullet points were captured in controlled environments, i.e., computer vision laboratories where the photos are treated artificially: constant backgrounds, homogeneous luminosity, and other conditions not usually occurring in lemon crops. Our dataset stands out from the others because the images were captured during cloudy, rainy, sunny, and windy days and considered scenarios with a variety of backgrounds in a typical lemon crop. This ensures that the algorithms learn from representative images of the type and complexity of real-world scenes.

Author Contributions: J.P.-A.: conceptualization, methodology, investigation, resources, data curation, writing—original draft, writing—review. R.A.-C.: data curation, investigation, writing—original draft, writing—review. J.M.C.: writing—review. M.C.: data curation. S.A.: data curation. A.L.: data curation. F.M.: data curation. All authors have read and agreed to the published version of the manuscript.

Funding: This research received no external funding.

Data Availability Statement: Data regarding images and annotations can be accessed at: repository name: *LeLePhid*; data identification number: DOI: 10.17632/tndhs2zng4; direct URL to data: https://data.mendeley.com/datasets/tndhs2zng4; accessed date: 13 January 2021.

Conflicts of Interest: The authors declare no conflict of interest.

References

1. Chen, T.; Zeng, R.; Guo, W.; Hou, X.; Lan, Y.; Zhang, L. Detection of Stress in Cotton (*Gossypium hirsutum* L.) Caused by Aphids Using Leaf Level Hyperspectral Measurements. *Sensors* **2018**, *18*, 2798. [CrossRef] [PubMed]
2. Virginio-Filho, E.; Astorga, C. *Prevención y Control de la Roya del Café. Manual de Buenas Prácticas para Técnicos y Facilitadores*; Centro Agronómico Tropical de Investigación y Enseñanza (CATIE): Turrialba, Costa Rica, 2015.
3. Suo, X.; Liu, Z.; Sun, L.; Wang, J.; Zhao, Y. Aphid Identification and Counting Based on Smartphone and Machine Vision. *J. Sens.* **2017**, *2017*, 3964376:1–3964376:7.
4. Parraga-Alava, J.; Cusme, K.; Loor, A.; Santander, E. RoCoLe: A robusta coffee leaf images dataset for evaluation of machine learning based methods in plant diseases recognition. *Data Brief* **2019**, *25*, 104414. [CrossRef] [PubMed]
5. McHugh, M.L. Interrater reliability: The kappa statistic. *Biochem. Medica* **2012**, *22*, 276–282. [CrossRef]
6. Parraga-Alava, J.; Alcivar-Cevallos, R.; Riascos, J.A.; Becerra, M.A. Aphids Detection on Lemons Leaf Image Using Convolutional Neural Networks. In *Systems and Information Sciences*; Botto-Tobar, M., Zamora, W., Larrea Plúa, J., Bazurto Roldan, J., Santamaría Philco, A., Eds.; Springer International Publishing: Cham, Switzerland, 2021; pp. 16–27.
7. Chen, J.; Fan, Y.; Wang, T.; Zhang, C.; Qiu, Z.; He, Y. Automatic Segmentation and Counting of Aphid Nymphs on Leaves Using Convolutional Neural Networks. *Agronomy* **2018**, *8*, 129. [CrossRef]
8. Bah, M.D.; Hafiane, A.; Canals, R. Deep Learning with Unsupervised Data Labeling for Weed Detection in Line Crops in UAV Images. *Remote Sens.* **2018**, *10*, 1690. [CrossRef]
9. Cālina, J.; Cālina, A.; Miluţ, M.; Croitoru, A.; Stan, I.; Buzatu, C. Use of drones in cadastral works and precision works in silviculture and agriculture. *Rom. Agric. Res.* **2020**, *37*, 273–284.
10. Kitano, B.T.; Mendes, C.C.T.; Geus, A.R.; Oliveira, H.C.; Souza, J.R. Corn Plant Counting Using Deep Learning and UAV Images. *IEEE Geosci. Remote Sens. Lett.* **2019**, 1–5. [CrossRef]
11. Gomez Selvaraj, M.; Vergara, A.; Montenegro, F.; Alonso Ruiz, H.; Safari, N.; Raymaekers, D.; Ocimati, W.; Ntamwira, J.; Tits, L.; Omondi, A.B.; et al. Detection of banana plants and their major diseases through aerial images and machine learning methods: A case study in DR Congo and Republic of Benin. *ISPRS J. Photogramm. Remote. Sens.* **2020**, *169*, 110–124. [CrossRef]

 *data*

Article

Distinct Two-Stream Convolutional Networks for Human Action Recognition in Videos Using Segment-Based Temporal Modeling

Ashok Sarabu * and **Ajit Kumar Santra ***

SITE, VIT University, Vellore, Tamil Nadu 632014, India
* Correspondence: sarabu.ashok@gmail.com (A.S.); ajitkumar@vit.ac.in (A.K.S.)

Received: 16 October 2020; Accepted: 8 November 2020; Published: 11 November 2020

Abstract: The Two-stream convolution neural network (CNN) has proven a great success in action recognition in videos. The main idea is to train the two CNNs in order to learn spatial and temporal features separately, and two scores are combined to obtain final scores. In the literature, we observed that most of the methods use similar CNNs for two streams. In this paper, we design a two-stream CNN architecture with different CNNs for the two streams to learn spatial and temporal features. Temporal Segment Networks (TSN) is applied in order to retrieve long-range temporal features, and to differentiate the similar type of sub-action in videos. Data augmentation techniques are employed to prevent over-fitting. Advanced cross-modal pre-training is discussed and introduced to the proposed architecture in order to enhance the accuracy of action recognition. The proposed two-stream model is evaluated on two challenging action recognition datasets: HMDB-51 and UCF-101. The findings of the proposed architecture shows the significant performance increase and it outperforms the existing methods.

Keywords: segment-based temporal modeling; two-stream network; action recognition

1. Introduction

Human Action Recognition is an emerging research area that has gained prominent attention in computer vision. One of the reason for which researchers are interested in the action recognition in video is the wide range of its applications in human-computer communication, video retrieval, management of web videos, surveillance [1], medicine, etc. When ompared to the still image recognition, temporal content in the video provides supplemental data for action recognition, as the number of actions can be accurately recognized using motion information. Action recognition in videos is a strenuous job because of the similarity in visual contents (frames) [2], view-point variation, camera motion, scale and pose of actor, and lighting conditions. Recently, the introduction of deep CNNs has made a major breakthrough performance in speech and image recognition tasks. Since then, computer vision researchers have started to apply the deep CNNs to action recognition in videos [3,4].

Deep learning in video action recognition is relatively slow when compared to image recognition. There are two reasons; first, scale and diversity of the video action recognition datasets are relatively small as compared to the image recognition datasets. Thus, small datasets will lead to overfitting, and the model will not be generalized for recognition. It is hard to create large-scale video datasets and train them on depth networks. Second, when compared to the image datasets, video data will contain an additional cue, called temporal information, which needs complex data analysis. Recently, many researchers have made attempts to solve these challenges and proposed solutions. Karpathy et al. [3], studied the performance of video action recognition on SPORTS-1M video classification dataset, compared the different CNN models. Du et al. proposed the C3D model, a three-dimensional convolution network for video action recognition. Later, Simonyan et al. [5] presented a two-stream architecture for the first time for action recognition in videos, that works on two CNNs which showed good performance improvement. The researchers for the aforementioned methods are able to utilize the temporal component, but work only for a short time; in lengthy videos, information cannot persist for a long time. To solve this problem, Wang et al. [6] designed a video level segmental architecture, called Temporal Segment Networks that can efficiently learns the features and retrieve the long-range time-varying features from the videos.

In this paper, we propose a two-stream CNN model for identifying actions in videos built on a two-stream network model. The proposed architecture is inspired from a two-stream idea [5], a two-stream model with the similar two-stream structures for human action recognition in videos. Specifically, the RGB image is the input to the spatial stream. Furthermore, the stack of consecutive optical flow images is the input to the temporal stream. Each stream is implemented while using identical two-stream, and the final results of both streams are combined with the late fusion technique. The other methods proposed in [5–11], by researchers utilized similar network models for two streams for human action recognition in videos. However, in human visual cortex systems, recognizing an object and its action are entirely two different processes. Inspired by the human visual cortex process, we proposed similar two-stream CNN architecture for action recognition in videos. Because of the variable length of videos, we attempt to add a video segmentation technique [6], to retrieve the long term temporal features. The proposed model of two-stream convolutional network is shown in Figure 1. Data augmentation and advanced cross-modal pretraining are employed because of the small size of datasets and to avoid labeled noise. The first step in our model is segmenting video in three parts. In the next step, the three snippets are randomly sampled and fed into the proposed two-stream network. Subsequently, the final category score is captured at the end of each stream of the network and fused for the final video level prediction.

With experimental results of our proposed model using the two most popular action recognition datasets, HMDB-51, and UCF-101, the contribution to this papers is three-fold. First, two-stream with multiple networks produces better performance than the two-stream with similar network models. In our experimentation, we found that ResNets and Inception-V2 produced better feature extraction and performance than other network models. Second, data augmentation and advanced cross-modal pre-training techniques are employed because of existing small datasets and noisy labels. Finally, the segment based temporal modeling technique for long-term temporal information better captures long-range information.

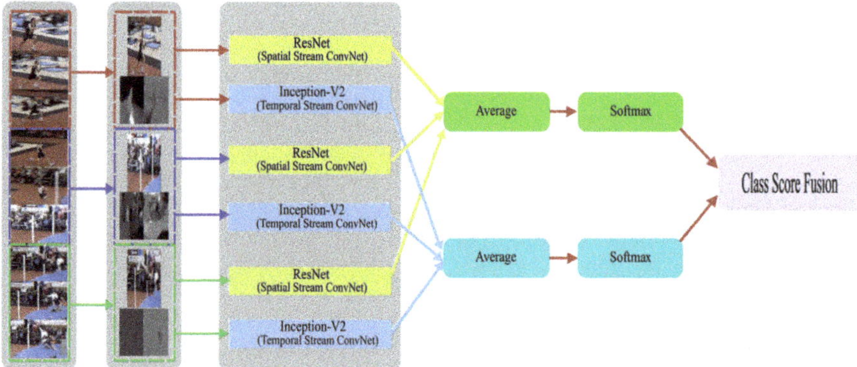

Figure 1. Distinct Two-Stream Convolutional Networks for Human Action Recognition in Videos while using Segment-Based Temporal Modeling.

2. Related Works

Recently, deep CNNs attained tremendous success in image recognition. Driven with the success of CNNs in image recognition, computer vision researchers transferred to videos. Action recognition in videos in deep learning is categorised into three categories that are based on the network architectures (1) Space-time networks. (2) Hybrid networks. (3) Two-stream networks.

2.1. Space-Time Networks

Space-time networks are the two-dimensional convolution networks with an additional convolution operation for temporal information. Ji et al. [12] presented a model that is one of the seminal works, recognizes actions in videos applying convolution neural networks. Ji et al. [12] extract spatial and temporal information by applying three-dimensional convolutions on adjacent frames. The networks repeat the same three-dimensional (3D) convolutions and sampling. Finally, a 128-dimensional feature vector is generated and it is used for action classification.

The 3D CNN in [12] was later extended to three-dimensional convolution networks [4], a deep network architecture trained on large scale datasets. The three-dimensional convolution networks contain five convolution layers, five max-pooling layers, two fully connected layers, and a softmax loss function layers. Even though information of the two streams is considered in training, the overall cost of computing and model storage is remarkably high. Liu et al. [13] proposed SSNET, stack of convolutional layers are added to temporal data and showed the best performance on skeleton data. Diba et al. [14] presented a three-dimensional temporal architecture, a new temporal layer called "Temporal Transition Layer" applied in the 3D DenseNet-based network. This method ignored the temporal information and only evaluated the RGB frames. Qui et al. [15], proposed Pseudo-3D Residual Network, (3*3*3) convolution filters are replaced with (1*3*3) in the spatial stream and (3*1*1) convolution filters in the temporal stream. (3*1*1) convolution filter is used in order to extract spatial information, and (1*3*3) is used to retrieve temporal features. When compared to the 3D convolution networks, this architecture is successful in terms of performance in video action recognition.

2.2. Hybrid Networks

Hybrid networks work on the principle of aggregating temporal information [16,17]. The aggregation of temporal information is done by adding the recurrent layers on the top layers of CNN's. These networks take advantage of both CNNs and LSTMs, and shows the positive results in capturing the spatial information, temporal information, and long-range dependencies [8,18–20]. Wang et al. [16], presented Long-term Recurrent Convolution Network (LRCN), in which frames were processed with CNN, and the output of CNN is fed into a stack of LSTMs. Veeriah et al. [21], designed a different gating for LSTM, called differential Recurrent Neural Network (dRNN). This method is good at learning significant spatio-temporal structures. Ng et al. [17] proposed two methods that can handle full-length videos. Two methods aggregates frame-level outputs of CNN to video level prediction. They discussed six different methods that showed, adding of LSTM layers after CNN outperforms temporal pooling information. Wu et al. [22] presented a hybrid network that uses both CNN and LSTM. They first extract spatial, temporal features with CNN and later fed as input to LSTM network for long term temporal features. However, because of the additional parameter of LSTM, LSTM has not shown the acceleration in performance in action recognition in videos.

2.3. Two-Stream Networks

Two stream networks use two CNNs for spatial and temporal information in videos. Simonyan et al. [5], presented a two-stream architecture for action recognition in videos. In this architecture, RGB images are fed into a spatial stream, optical flow frames are fed into temporal stream. Finally, softmax scores are obtained by fusing outputs of two-stream outputs. Wang et al. [8] presented Trajectory-pooled Deep-convolution Descriptor and integrated trajectory features and deep network learned features. This method shows superior performance by combining deep networks and shallow local features. Feichtenhofer et al. [9] proposed a new spatiotemporal architecture and explored various fusing schemes. They found fusing network spatially at last convolution layers boosts accuracy. Wang et al. [6] introduced Temporal Segment Networks, in which they improved the performance by training on the whole video by modeling long-range temporal structure. Yunbo et al. [11] introduced an end-to-end learning neural network, which combinedly performs pixel level action recognition and segmentation. They solved the action recognition by two-stream network along with temporal aggregation. Christoph et al. [7], designed a spatio-temporal ResNet, which allows the learning of the spatio-temporal feature by connecting static and optical flow channel streams, which increases the interactions between both streams.

3. Technical Approach

In this section, we provide a comprehensive overview of our proposed network architecture for action recognition in videos. First, we discuss the proposed distinct two-stream convolutional networks for human action recognition in videos. Subsequently, ResNet and Inception-V2 used as CNN model for spatial stream and temporal stream are discussed. After that, Temporal segmentation Network is introduced to capture the long-range temporal features. Finally, optimizing the network training strategies are presented.

3.1. Distinct Two-Stream Convolution Networks

Video is a collection of spatial and temporal information. The human visual cortex system that is mentioned in [23] processes information with two streams called spatial stream and temporal stream. Information is static image appearance in spatial stream; it only depicts scenes and objects. In the temporal stream, information is the movement of objects between consecutive frames, conveys the orientation of camera and objects. Inspired from [5,23], designed a two-stream CNN to retrieve the spatial and temporal features with two similar CNN models from videos. For spatial information, RGB frames are used. For temporal information, dense optical flow frames are used in order to extract the motion of objects across the video. Each of these streams is processed identical and independent deep convolutional neural network models. Specifically, the RGB image is fed to the spatial stream CNN. For temporal stream, stack of optical flow images are fed as input. Optical flow images are a combination of horizontal and vertical convoluted images. The number of optical flow images (L) is set to ten. Because the optical flow images consist of both horizontal and vertical convoluted images, the total number of flow images is set to 2L = 20 [5]. Finally, spatial and temporal streams are individually trained end-to-end, and the output of two streams are combined to get the final classification decision. Averaging and SVM are two fusing methods used in [5], in order to fuse scores of two streams.

In this subsection, we present our distinct two-stream CNN for action recognition based on the architecture presented in [5]. In this network architecture, the spatial and temporal streams are trained with different CNN models, as shown in Figure 2. The reasons for modeling our proposed architecture is, when two-streams with similar CNNs are trained and fused together, generates a large number of redundant features. Because optical flow frames are horizontal and vertical components that are derived from the RGB image, and when trained with a similar CNN generates redundant features. The second reason is, in human action recognition, object recognition and motion recognition are two different processes. Similarly, here. two-stream action recognition can be trained with two different CNN models. With many experiments, we observed that the performance of distinct two-stream action recognition is better than the two-stream model with similar CNNs.

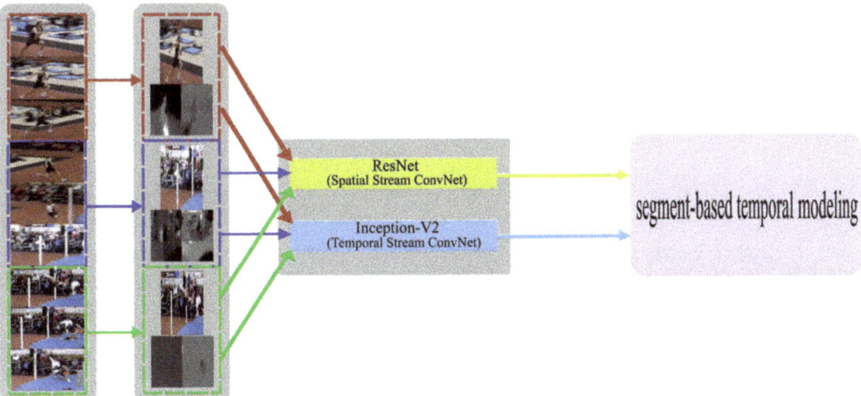

Figure 2. Distinct Two-Stream Convolutional Network.

3.2. Base Networks

In the above section, we discussed a two-stream spatio-temporal action network. A good action recognition model will retrieve more discrete spatial and temporal features. Previous studies [24,25] have shown that deeper CNN models can extract discrete features. In [25], the features of hidden layers and its mechanism of the CNN model were visualized. Moreover, when compared to the different CNN models with different depths and going deeper layer, the CNN model is better in extracting discriminant features, which can increase the prediction rate of the model. Another set of recent studies [26,27], showed that with the increase in network depth can learn more features in lengthy videos. Residual Networks (ResNet) in [24,28], addressed the issue of degradation [29], which is caused by deep layers of the CNNs. ResNets and Inception-V2 are the underlying networks in our models. ResNet is used to retrieve spatial and temporal features and Inception-V2 is utilized to increase the performance of model. We introduce two models ResNet and Inception-V2, in order to investigate them further and explore the potential of the distinct two-stream CNN.

3.2.1. Residual Network

ResNet is used as one of the CNN models in our proposed network. The primary reason to use ResNet with deep layers is that it extracts discriminant features from frames. Network degradation arises as to the number of layers increases. To solve this issue, He et al. [24] presented a deep ResNet. Instead of the original underlying fitting function, they use a trained residual network by residual unit is

$$x_{i+1} = \sigma\left(x_i + F\left(x_i; W_i\right)\right) \tag{1}$$

where x_i and x_{i+1} are the input and output of the ith layer of the network. $F\left(x_i; W_i\right)$ is non-linear residual mapping of the weight of CNN filters $W_l = \left\{W_{l,k}\big|_{1<k<K}\right\}$, and sigma is the ReLU function [30]. The benefit of using the residual block is that it acts as the shortcut connection that connects the first layer to any layer in the network, which breaks the conventional form of connecting one layer to the next layer. With this, the gradient loss may skip some layers and pass from the loss layer to any layer that it is connected, and this will avoid the gradient explosion problem. This shortcut connection does not increase the computational cost and an increase in the number of parameters. In ResNet, after every convolution operation and before the activation layer, batch normalization (BN) [31], is performed. This will solve the covariate shift problem and, also, the convergence of the network will be fast [24]. Finally, the global average pooling and softmax layer are employed combinedly in the place of a single fully connected layer. This effectively decreases the number of

parameters. Besides, the bottleneck structure will decrease the computational overhead, and the network efficiency is guaranteed.

3.2.2. Inception-V2

Inception-V2 [31] is used as another CNN model in our proposed network. Inception-V2 is the module used in order to reduce the CNN complexity. This CNN model is the advanced version of the GoogleLeNet [26], which solves the saturation and vanishing gradient problem. The distribution of X is to be unchanged, because even a minor change will be change the value of X when the network goes deep. A higher learning rate can be used for faster optimization. The primary concept of the Inception-V2 is the replacement of 5*5 convolution with two 3*3 convolutions. This replacement of convolution will not only decrease the parameter number, but also increase more non-linear transformations, enhances the model to learn more features. Other advantages of adding the batch normalization reduce the internal covariate shift by normalizing the output of each layer to N(0,1).

In original two-stream CNN for video action recognition [5], VGG-M-2048 [32], is used to train the model, and both of the streams use the same network structure. Feichtenhofer et al. [9], improved the performance using VGG-16 [25] instead of VGG-M-2048. ResNet and Inception-V2 are used as the CNN models in our proposed architecture. ResNets with an increase in the number of layers can extract more features [24]. Furthermore, Inception-V2, with an increase in the network depth and width, can improve performance [31]. Looking at the benefits of the ResNet and Inception-V2, we used these as base models in our two-stream architecture. ResNet has less computational complexity and filters when compared to the VGG-M-2048. In terms of computational complexity, VGG-16 uses 15.3 B FLOPs and VGG-19 uses 19.6 B FLOPs, whereas ResNet-152 only uses 11.3 B FLOPs. Similarly, the computational complexity of ResNet-50 is 3.8 B FLOPs and ResNet-101 is 7.6 B FLOPs. Finally, the total number of parameters of both streams are 182 M in our model.

3.3. Segment-Based Temporal Modeling

Problem with the original two-stream CNN [5] architecture is its inability to maintain temporal information in deep CNN networks. The cause for this problem is, it only works on one frame in the spatial stream or a stack of optical flow frames for the temporal stream. Therefore, the network is unable to retrieve long-range temporal information effectively. Segment based long-range temporal information plays an important role in finding action recognition in videos. For example [22,33], in some complex video actions, comprises multiple stages are required in order to classify the action and subject. And, action is important from the beginning to the final point of the video (basketball dunk and shooting similar for some short time, so start to the endpoint to be considered to classify correct action). Therefore, there may be misclassification of action if a video is considered only for some part of the time, which leads to unsatisfied performance. To improve the performance, we implement the long-range temporal model proposed in [6] to extract long-term temporal information in our proposed distinct two-stream convolutional networks for human action recognition in videos.

In order to model the long-range segment based temporal modeling [6], we divided the video into K segments (K = 3) in equal duration, expressed as {S1, S2, S3}. For short snippets, modeling is done while using,

$$TSN\left(T_1, T_2, \cdots, T_j\right) = \mathcal{H}\left(\mathcal{G}\left(\mathcal{F}\left(T_1; \mathbf{W}\right), \mathcal{F}\left(T_2; \mathbf{W}\right), \cdots, \mathcal{F}\left(T_j; \mathbf{W}\right)\right)\right) \tag{2}$$

where $\mathcal{F}(T_1; \mathbf{W})$ represents Convolutional function with parameters, $\mathcal{G}$ represents an averaging function, H represents softmax function. Subsequently, we sample each segment (S_j) into short snippets $\{T_1, T_2, T_3\}$. These short snippets are fed as an input to the proposed two-stream architecture to get an initial action classification score. Afterwards, this score is fused with average function to obtain a final decision among snippets. Based on this consensus, the final prediction scores are calculated while using the widely used softmax function. Additionally, the final loss function of segmental consensus is calculated while using the equation,

$$\mathcal{L}(y, \mathbf{G}) = -\sum_{i=1}^{C} y_i \left(G_i - \log \sum_{j=1}^{C} \exp G_j \right) \tag{3}$$

where 'n' is the total number of the acting categories, y_i is the ground-truth label, $\mathbf{G}$ is classification score of i. The value of $\mathbf{G}$ is average result of the short snippets of same categories. For the proposed architecture, all segmental frames are utilized together to optimize the network parameter $\mathbf{W}$. In the backpropagation, the gradient of W to the loss value L can be derived as,

$$\frac{\partial \mathcal{L}(y, \mathbf{G})}{\partial \mathbf{W}} = \frac{\partial \mathcal{L}}{\partial \mathbf{G}} \sum_{k=1}^{K} \frac{\partial \mathcal{G}}{\partial \mathcal{F}(T_k)} \frac{\partial \mathcal{F}(T_k)}{\partial \mathbf{W}} \tag{4}$$

Subsequently, we use stochastic gradient descent (SGD) to train the model parameters. As in Equation (4), guarantees that the model parameters are updated using the segmental consensus category $\mathbf{G}$ for three short snippets. Thus, with this optimization technique, long-range segment based temporal information is preserved, and model parameters are learned from the entire video. Aggregation function ($\mathbf{G}$) is the main concept in segment-based temporal modeling. The averaging function is an aggregation function used to predict by averaging final results at the snippet level for every class with $g_i = \frac{1}{N} \sum_{n=1}^{N} f_i^n$. Final result of this function, g_i with respect to f_i^k is,

$$\frac{\partial g_i}{\partial f_i^n} = \frac{1}{N} \tag{5}$$

4. Network Training Strategies

4.1. Data Augmentation

Data augmentation is used in our model in order to create manifold training samples. Data augmentation strategies are used when there are fewer training samples to avoid overfitting problems. In our proposed model, horizontal flipping, random cropping, and scale-jittering [27] are employed to augment the training data. In corner cropping, the extracted regions are only selected from the corner or center of the image to avoid focus of the information only on the center of the image. We use the multi-scale jittering technique [34], which is applied in ImageNet classification. We set the input size of the image or optical flow field to 256 × 340. Height and width of cropped region are selected randomly from {256,224,192,168} and resized to 224 × 224 for model training.

4.2. Advanced Cross-Modal Pre-Training

Pre-training is a great way to initialize the deep convolution neural network model when the dataset size is small, i.e., when the training samples are less [5]. Input to the Spatial stream network is RGB images, so a pre-trained model can be used, such as ImageNet [30], in order to set the initial weights of CNN. However, the input to the temporal stream network is optical flow frames, and, Optical flow frames and RGB difference contain the distinct features of video, and their data distribution is not the same as RGB images. Therefore, it is not possible to use pre-trained networks for temporal stream networks. Accordingly, we propose an advanced cross-modal pre-training technique. First, we apply the linear transformation operation [6] on optical flow frames to get the values in an interval of [0, 255]. Now, the values of optical flow frames will be in the same range of RGB images. Then, the first layers of CNN weights of RGB models are modified to fit the weights of the optical flow fields (because the RGB image has three channels and temporal stream input has 10 inputs, including horizontal and vertical images, we average the weight of the three channels weights of RGB to replicate the channel number of temporal network input (output kernel size = (64,10,7,7))). We do this process from scratch, and then we replace the values of the first layer of CNN model with the values of same layers of the RGB pre-trained model.

5. Experiments

In this section, we discuss the implementation details of proposed architecture and the datasets. Subsequently, we evaluate the performance of two-stream networks with similar and distinct network architectures. Afterwards, the performance of the proposed advance cross-modal pre-training is presented. Finally, we present the experimental results and analysis in the last section.

5.1. Datasets and Implementation Detials

We perform experiments on large-scale action recognition datasets, namely UCF101 [35] and HMDB51 [36]. The UCF101 dataset consists of 101 action classes with 13,320 videos in total. Each video consists of an average of 100–300 frames with a duration of 3–10 s. The HMDB51 dataset consists of video clips from different online sources, such as YouTube and Google. The dataset consists of 51 action categories with 6766 videos in total. We evaluate the proposed two-stream architecture by following the standard evaluation scheme while using three training and testing splits of the UCF-101 dataset. Affitionally, the results compared with state-of-art methods. The evaluation of average accuracy is made on three splits of UCF-101 and HMDB-51.

The mini-batch gradient descent method is implemented to train the network parameters. We initialize the network parameters with pre-training models from ImageNet [30]. We initialize the values of batch size, weight decay, and momentum to 256, 0.0005, and 0.9, respectively. Initially, both stream's learning rate is initialized to 10^{-4}. When training the spatial stream network learning rate is decreased to 10^{-1} for every 15 K iterations and the entire network training halts at 36 K iterations. Similarly, when training the temporal stream network, the learning rate is decreased to $1/10$ at 20 K and 32 K iterations and the entire network training halts at 40 K iterations. TVL1 optical flow algorithm [37] is used to extract optical flow frames from videos. To speed-up the training process, we apply data-parallelization with multiple GPUs on the Caffe platform [38] and related code is released on GitHub (https://github.com/ashoksarabu/Distinct-Two-Stream).

5.2. Testing

We evaluate our proposed model with the parameters of the original two-stream convolution network [5]. We sample a fixed number of RGB images or optical flow stacks (25 in our experiment) with an equal interval of times between them. For each of the frames, we crop four corners, one center, and horizontal flipping to evaluate the CNNs. We use weighted averaging to fuse two stream's results. When the network is trained, the performance gap between two streams is smaller than the original two-stream convolution network [5]. Because of this small gap, we initialize the weights of spatial stream to 1 and temporal stream to 1.5.

5.3. Exploration Study

In this section, we examine and evaluate the efficiency of the proposed network with the two-stream identical networks. We propose an improved cross-modal pre-training approach in Section 3 is evaluated in the experiments and its effectiveness of the proposed network model.

The experimental tests are performed on the proposed CNN architecture using the same CNN model with different depths, and with different CNN models. Inception-V2 [31] and ResNet with different depths are used to evaluate and test the model. ResNet-50, ResNet-101, and ResNet-152 [24] are ResNets with different depths used as CNN for both streams. The experimental results of the proposed model are evaluated and compared in Table 1. The comparisons of experimental results are made based on 1. Two streams with the identical CNN model, 2. Two streams with non-identical network models and depths. From Table 1, we found out that ResNet-101 performed better for spatial stream network. When compared to the two streams CNN with similar networks, two-stream CNNs with different networks performed better.

Table 1. Performance of Distincttwo-stream Convolutional Network on UCF-101.

Network Architectures for Two-Streams	Spatial	Temporal	Two-Stream
Spatial_ResNet-101 + Temporal_ResNet-50	84.1%	85.3%	94.3%
Spatial_ResNet-152 + Temporal_ResNet-50	86.2%	85.3%	94.3%
Spatial_ResNet-101 + Temporal_Inception-V2	84.1%	88.8%	95.0%

Moreover, similar networks with different depths performed well as compared to similar networks with similar depths. ResNet-50 performed better for the temporal stream and ResNet-101 for the spatial stream network. We achieved the best performance with an accuracy of 95.00 percent when ResNet-101 is used as the spatial stream network model, and Inception-V2 is used as the temporal stream network model.

We evaluate the experiments with ResNet-50 and Inception-V2 models to verify the efficiency of advanced cross-modal pre-training technique discussed in the previous section, as mentioned above. Specifically, three case-studies are used. First, training the temporal stream network from scratch. Second, training the temporal stream network with the technique proposed in [6]. Third, training the temporal stream network with our proposed method. The experimental results of the three case studies mentioned earlier are performed on UCF-101 dataset and tabulated in Table 2. From the results that are tabulated in Table 2, we summarize that method used for pre-train the temporal stream network and initializing a deep convolution network achieved great accuracy when compared to training from scratch. Moreover, the proposed advanced cross-modal pre-trained has an increase of 0.3% with CNN models ResNet-50 and Inception-V2 when compared to the method proposed in [6].

Table 2. Performance evaluation of temporal stream CNN on UCF-101dataset.

Training Strategy	ResNet-50	Inception-V2
From scratch	78.5%	82.4%
Pre-Training [6]	84.1%	87.3%
Proposed - Advanced cross-modal pre-training	85.3%	88.8%

5.4. Comparison with State-of-the-Art

After investigating the different models for two-stream models for recognizing human action in videos, we found the optimal accuracy. We evaluation of the proposed model are based on the UCF-101 and HMDB-51 action recognition datasets on all splits and reported. The empirical results are presented in Table 3. When compared to state-of-the-art results, our proposed architecture with ResNet-101 for the spatial stream network and the Inception-V2 model for the temporal stream network, has performed better. Compared to the original two-stream convolution neural network [5] and ST-ResNet model [7], the accuracy for the UCF-101 dataset has been improved by 7.10% and 1.7%. Similarly, for the HMDB-51 dataset, compared to the original two-stream convolution neural network [5] and the ST-ResNet model [7], we got optimal accuracy has been increased by 8.5% and 1.5%, respectively. From the experimental findings, we conclude that effectiveness of our distinct two-stream convolutional network for human action recognition in videos based on segment based temporal modeling. Furthermore, spatiotemporal heterogenous network accuracy has been improved compared to the two-stream action recognition methods with similar network models.

Table 3. Comparison of our proposed method distinct two-stream convolutional network with state-of-art methods on UCF-101 andHMDB-51 datasets.

Methodology	UCF-101	HMDB-51
Two-stream network [5]	88.0%	59.4%
Two-stream network fusion [9]	92.5%	65.4%
Spatio-Temporal 3D CNNs [4]	85.2%	–
Factorized Spatio-Temporal CNNs [36]	88.1%	59.1%
Pseudo-3D residual networks [14]	93.7%	–
Temporal Segment Networks [6]	94.0%	68.5%
Temporal 3D CNNs [13]	93.2%	63.5%
SpatioTemporal residual networks [7]	93.4%	66.4%
(Proposed) Distinct two-stream CNN	95.0%	67.9%

6. Conclusions

In this paper, we presented a distinct two-stream convolutional networks for recognizing human action in videos using segment based temporal modeling. Human action recognition is two individual processes, which is, two different independent streams processes appearance and motion. Inspired by this, we attempted to experiment with the two-stream convolution neural network with two different network models for two streams. Additionally, we achieved the best performance when compared to existing two-stream networks. With all the experiments, it is found that the distinct two-stream convolution networks for recognizing action in videos perform better than two-stream convolution networks with similar network models. In our experiments, we found that ResNet-101 and Inception-V2 models, when employed as network models for a two-stream network with segment based temporal modeling, yield the best performance. Finally, data augment techniques and advanced cross-modal pretraining are applied in order to increase the performance.

Author Contributions: Conceptualization, A.S.; Investigation, A.S. and A.K.S.; Methodology, A.S.; Resources, A.S.; Software, A.S.; Supervision, A.K.S.; Validation, A.K.S.; Writing—original draft, A.S.; Writing—review & editing, A.S. and A.K.S. All authors have read and agreed to the published version of the manuscript.

Funding: This research received no external funding.

Conflicts of Interest: The authors declare no conflict of interest.

References

1. Nanda, A.; Sa, P.K.; Choudhury, S.K.; Bakshi, S.; Majhi, B. A neuromorphic person re-identification framework for video surveillance. *IEEE Access* **2017**, *5*, 6471–6482. [CrossRef]
2. Nanda, A.; Chauhan, D.S.; Sa, P.K.; Bakshi, S. Illumination and scale invariant relevant visual features with hypergraph-based learning for multi-shot person re-identification. *Multimed. Tools Appl.* **2019**, *78*, 3885–3910. [CrossRef]
3. Karpathy, A.; Toderici, G.; Shetty, S.; Leung, T.; Sukthankar, R.; Fei-Fei, L. Large-scale video classification with convolutional neural networks. In Proceedings of the IEEE Conference on Computer Vision and Pattern Recognition, Columbus, OH, USA, 24–27 June 2014; pp. 1725–1732
4. Tran, D.; Bourdev, L.; Fergus, R.; Torresani, L.; Paluri, M. Learning spatiotemporal features with 3d convolutional networks. In Proceedings of the IEEE International Conference on Computer Vision, Santiago, Chile, 7–13 December 2015; pp. 4489–4497.
5. Simonyan, K.; Zisserman, A. Two-stream convolutional networks for action recognition in videos. In Proceedings of the Advances in Neural Information Processing Systems, Montreal, QC, Canada, 8–13 December 2014; pp. 568–576.
6. Wang, L.; Xiong, Y.; Wang, Z.; Qiao, Y.; Lin, D.; Tang, X.; Gool, L.V. Temporal segment networks: Towards good practices for deep action recognition. In *European Conference on Computer Vision*; Springer: Cham, Switzerland, 2016; pp. 20–36.

7. Christoph, R.P.W.; Pinz, F.A. Spatiotemporal residual networks for video action recognition. In Proceedings of the Advances in Neural Information Processing Systems, Barcelona, Spain, 5–10 December 2016; pp. 3468–3476.

8. Wang, L.; Qiao, Y.; Tang, X. Action recognition with trajectory-pooled deep-convolutional descriptors. In Proceedings of the IEEE Conference on Computer Vision and Pattern Recognition, Boston, MA, USA, 7–12 June 2015; pp. 4305–4314.

9. Feichtenhofer, C.; Pinz, A.; Zisserman, A. Convolutional two-stream network fusion for video action recognition. In Proceedings of the IEEE Conference on Computer Vision and Pattern Recognition, Las Vegas, NV, USA, 27–30 June 2016; pp. 1933–1941.

10. Ji, J.; Buch, S.; Soto, A.; Niebles, J.C. End-to-end joint semantic segmentation of actors and actions in video. In Proceedings of the European Conference on Computer Vision (ECCV), Munich, Germany, 8–14 September 2018; pp. 702–717.

11. Wang, Y.; Long, M.; Wang, J.; Yu, P.S. Spatiotemporal pyramid network for video action recognition. In Proceedings of the IEEE Conference on Computer Vision and Pattern Recognition, Honolulu, HI, USA, 21–26 July 2017; pp. 1529–1538.

12. Ji, S.; Xu, W.; Yang, M.; Yu, K. 3D convolutional neural networks for human action recognition. *IEEE Trans. Pattern Anal. Mach. Intell.* **2012**, *35*, 221–231. [CrossRef] [PubMed]

13. Liu, J.; Shahroudy, A.; Wang, G.; Duan, L.Y.; Kot, A.C. Skeleton-based online action prediction using scale selection network. *IEEE Trans. Pattern Anal. Mach. Intell.* **2019**, *42*, 1453–1467. [CrossRef] [PubMed]

14. Diba, A.; Fayyaz, M.; Sharma, V.; Karami, A.H.; Arzani, M.M.; Yousefzadeh, R.; Gool, L.V. Temporal 3d convnets: New architecture and transfer learning for video classification. *arXiv* **2017**, arXiv:1711.08200.

15. Qiu, Z.; Yao, T.; Mei, T. Learning spatio-temporal representation with pseudo-3d residual networks. In Proceedings of the IEEE International Conference on Computer Vision, Venice, Italy, 22–29 October 2017; pp. 5533–5541.

16. Wang, K.; Wang, X.; Lin, L.; Wang, M.; Zuo, W. 3d human activity recognition with reconfigurable convolutional neural networks. In Proceedings of the 22nd ACM International Conference on Multimedia, Orlando, FL, USA, 3 November 2014; pp. 97–106.

17. Ng, J.Y.H.; Hausknecht, M.; Vijayanarasimhan, S.; Vinyals, O.; Monga, R.; Toderici, G. Beyond short snippets: Deep networks for video classification. In Proceedings of the IEEE Vonference on Computer Vision and Pattern Recognition, Boston, MA, USA, 7–12 June 2015; pp. 4694–4702.

18. Kar, A.; Rai, N.; Sikka, K.; Sharma, G. Adascan: Adaptive scan pooling in deep convolutional neural networks for human action recognition in videos. In Proceedings of the IEEE Conference on Computer Vision and Pattern Recognition, Honolulu, HI, USA, 21–27 July 2017; pp. 3376–3385.

19. Diba, A.; Sharma, V.; Gool, L.V. Deep temporal linear encoding networks. In Proceedings of the IEEE Conference on Computer Vision and Pattern Recognition, Honolulu, HI, USA, 21–26 July 2017; pp. 2329–2338.

20. Liu, J.; Shahroudy, A.; Perez, M.L.; Wang, G.; Duan, L.Y.; Chichung, A.K. Ntu rgb+ d 120: A large-scale benchmark for 3d human activity understanding. *IEEE Trans. Pattern Anal. Mach. Intell.* **2019**, *42*, 2684–2701. [CrossRef] [PubMed]

21. Veeriah, V.; Zhuang, N.; Qi, G.J. Differential recurrent neural networks for action recognition. In Proceedings of the IEEE International Conference on Computer Vision, Santiago, Chile, 7–13 December 2015; pp. 4041–4049.

22. Wu, Z.; Wang, X.; Jiang, Y.G.; Ye, H.; Xue, X. Modeling spatial-temporal clues in a hybrid deep learning framework for video classification. In Proceedings of the 23rd ACM International Conference on Multimedia, Brisbane, Australia, 13 October 2015; pp. 461–470.

23. Goodale, M.A.; Milner, A.D. *Separate Visual Pathways for Perception and Action*; Psychology Press: London, UK, 1992.

24. He, K.; Zhang, X.; Ren, S.; Sun, J. Deep residual learning for image recognition. In Proceedings of the IEEE Conference on Computer Vision and Pattern Recognition, Las Vegas, NV, USA, 27–30 June 2016; pp. 770–778.

25. Yu, W.; Yang, K.; Bai, Y.; Xiao, T.; Yao, H.; Rui, Y. Visualizing and comparing AlexNet and VGG using deconvolutional layers. In Proceedings of the 33 rd International Conference on Machine Learning, New York, NY, USA, 19–24 June 2016.

26. Szegedy, C.; Liu, W.; Jia, Y.; Sermanet, P.; Reed, S.; Anguelov, D.; Erhan, D.; Vanhoucke, V.; Rabinovich, A. Going deeper with convolutions. In Proceedings of the IEEE Conference on Computer Vision and Pattern Recognition, Boston, MA, USA, 7–12 June 2015; pp. 1–9.
27. Simonyan, K.; Zisserman, A. Very deep convolutional networks for large-scale image recognition. *arXiv* **2014**, arXiv:1409.1556.
28. He, K.; Zhang, X.; Ren, S.; Sun, J. Identity mappings in deep residual networks. In *European Conference on Computer Vision*; Springer: Cham, Switzerland, 2016; pp. 630–645.
29. Srivastava, R.K.; Greff, K.; Schmidhuber, J. Highway networks. *arXiv* **2015**, arXiv:1505.00387.
30. Krizhevsky, A.; Sutskever, I.; Hinton, G.E. Imagenet classification with deep convolutional neural networks. *Adv. Neural Inf. Process. Syst.* **2017**, *60*, 84–90.
31. Ioffe, S.; Szegedy, C. Batch normalization: Accelerating deep network training by reducing internal covariate shift. *arXiv* **2015**, arXiv:1502.03167.
32. Chatfield, K.; Simonyan, K.; Vedaldi, A.; Zisserman, A. Return of the devil in the details: Delving deep into convolutional nets. *arXiv* **2014**, arXiv:1405.3531.
33. Wang, L.; Qiao, Y.; Tang, X. Latent hierarchical model of temporal structure for complex activity classification. *IEEE Trans. Image Process* **2013**, *23*, 810–822. [CrossRef] [PubMed]
34. Vinyals, O.; Toshev, A.; Bengio, S.; Erhan, D. Show and tell: A neural image caption generator. In Proceedings of the IEEE Conference on Computer Vision and Pattern Recognition, Boston, MA, USA, 7–12 June 2015; pp. 3156–3164.
35. Soomro, K.; Zamir, A.R.; Shah, M. UCF101: A dataset of 101 human actions classes from videos in the wild. *arXiv* **2012**, arXiv:1212.0402.
36. Kuehne, H.; Jhuang, H.; Garrote, E.; Poggio, T.; Serre, T. HMDB: A large video database for human motion recognition. In Proceedings of the 2011 International Conference on Computer Vision, Barcelona, Spain, 6–13 November 2011; pp. 2556–2563.
37. Zach, C.; Pock, T.; Bischof, H. A duality based approach for realtime tv-l 1 optical flow. In *Joint Pattern Recognition Symposium*; Springer: Berlin/Heidelberg, Germany, 2007; pp. 214–223.
38. Jia, Y.; Shelhamer, E.; Donahue, J.; Karayev, S.; Long, J.; Girshick, R.; Guadarrama, S.; Darrell, T. Caffe: Convolutional architecture for fast feature embedding. In Proceedings of the 22nd ACM International Conference on Multimedia, Orlando, FL, USA, 3 November 2014; pp. 675–678.

Publisher's Note: MDPI stays neutral with regard to jurisdictional claims in published maps and institutional affiliations.

© 2020 by the authors. Licensee MDPI, Basel, Switzerland. This article is an open access article distributed under the terms and conditions of the Creative Commons Attribution (CC BY) license (http://creativecommons.org/licenses/by/4.0/).

Data Descriptor

A Data Descriptor for Black Tea Fermentation Dataset

Gibson Kimutai [1,2,*] [ID], **Alexander Ngenzi** [1] [ID], **Rutabayiro Ngoga Said** [1] [ID], **Rose C. Ramkat** [3] [ID] and **Anna Förster** [4] [ID]

1   African Center of Excellence in Internet of Things (ACEIoT), College of Science and Technology, University of Rwanda, P.O. Box, 3900 Kigali, Rwanda; yngenzi37@gmail.com (A.N.); said.rutabayiro.ngoga@gmail.com (R.N.S.)

2   Department of Mathematics, Physics and Computing, Moi University, P.O. Box, 3900-30100 Eldoret, Kenya

3   Department of Biological Sciences, Moi University, P.O. Box, 3900-30100 Eldoret, Kenya; rose.ramkat@mu.ac.ke

4   Sustainable Communication Networks, University of Bremen, 8359 Bremen, Germany; anna.foerster@comnets.uni-bremen.de

*   Correspondence: kimutaigibs@gmail.com

Abstract: Tea is currently the most popular beverage after water. Tea contributes to the livelihood of more than 10 million people globally. There are several categories of tea, but black tea is the most popular, accounting for about 78% of total tea consumption. Processing of black tea involves the following steps: plucking, withering, crushing, tearing and curling, fermentation, drying, sorting, and packaging. Fermentation is the most important step in determining the final quality of the processed tea. Fermentation is a time-bound process and it must take place under certain temperature and humidity conditions. During fermentation, tea color changes from green to coppery brown to signify the attainment of optimum fermentation levels. These parameters are currently manually monitored. At present, there is only one existing dataset on tea fermentation images. This study makes a tea fermentation dataset available, composed of tea fermentation conditions and tea fermentation images.

Dataset: 10.5281/zenodo.4469326.

Dataset License: Creative Commons Attribution 4.0 International.

Keywords: tea; fermentation; internet of things; detection; dataset

Citation: Kimutai, G.; Ngenzi, A.; Ngoga Said, R.; Ramkat, R.C.; Förster, A. A Data Descriptor for Black Tea Fermentation Dataset. *Data* **2021**, *6*, 34. https://doi.org/10.3390/data6030034

Academic Editors: Munish Kumar, R. K. Sharma and Ishwar Sethi

Received: 9 March 2021
Accepted: 15 March 2021
Published: 19 March 2021

Publisher's Note: MDPI stays neutral with regard to jurisdictional claims in published maps and institutional affiliations.

check for updates

Copyright: © 2021 by the authors. Licensee MDPI, Basel, Switzerland. This article is an open access article distributed under the terms and conditions of the Creative Commons Attribution (CC BY) license (https://creativecommons.org/licenses/by/4.0/).

1. Background and Rationale

Tea is currently among the most popularly consumed beverage across the world [1] and is responsible for the economic growth of many countries, including India, Sri-Lanka, Kenya, China among other countries [2]. These top tea-producing countries produce several varieties of tea, which include: yellow tea, illex tea, oolong tea, black tea, white tea, among others [3]. Among these categories of tea, black tea is the most consumed, accounting for approximately 78% of the total daily consumption of tea [4]. Kenya is the leading exporter of black tea worldwide, with her major tea-producing counties being Kericho, Bomet, Nandi, and Nyeri [5]. The crop is a source of livelihood for more than 10 million of the total countries' estimated population of 47 million people [6]. Although the crop is still the leading exchange earner for the county, the sector is ailing due to ever-reducing tea prices [7]. This is attributed to increased competition from other countries, poor management, and the low quality of tea produced, among others [8].

The steps of processing black tea are plucking, withering, crushing, tearing and curling, fermentation, drying, sorting, and packaging [9]. Among these processes, fermentation is the key determinant of the final quality of the processed tea [10]. The process is time-bound and must take place within a given temperature and humidity range [11]. During fermentation processes, tea changes color from green to coppery brown and finally to dark red [12]. The optimally fermented tea is coppery brown in color, and has a fruity smell and a sweet taste, while the unfermented tea is green in color, and has a grassy smell and a strong taste [13]. The overfermented tea is dark red and is characterized by a bitter taste [14]. Currently, the optimum fermentation of tea is monitored manually by tea tasters, adopting the following techniques: monitoring color change, tasting tea as fermentation progresses, and smelling the odor of tea during fermentation [15]. These manual methods are subjective and lead to a compromise in the quality of the made tea. Image processing and machine learning techniques have shown high levels of ability in various fields, including medicine [16], education, E-Commerce [17] tourism and banking, among others [18]. However, for image processing and machine learning to work, there is a need for data for training and evaluation of the models. Worryingly, there is only one reported open-source dataset on tea fermentation images [19], which is a limiting phenomenon as researchers have only one dataset to train and evaluate their machine learning models. Therefore, this study aims to resolve this challenge by adopting the Internet of Things (IoT) to capture and release a tea fermentation dataset composed of temperature, humidity, and black tea fermentation images.

The rest of the paper is arranged as follows. Section 3 presents the description of the data in the dataset. Section 2 presents materials and methods, and Section 5 presents the conclusion of the paper.

2. Materials and Methods

This section describes the resources and the approach followed in the collection of the dataset. Section 2.1 discusses on the resources while the collection of the dataset is discussed in Section 2.2.

2.1. Resources

The following resources were instrumental in acquiring the data: Raspberry pi, Pi-Camera, Server, and programming languages. Raspberry Pi model B+ was adopted due to its increased processing power and its dual-band Wi-Fi Feature [20]. The Raspbian operating system for raspberry pi [21] was used. Raspian was chosen as it is available at no cost and is easy to install and use. A raspberry pi camera of 8 megapixels was used. The board was chosen since it is very small, weighing around 3 g, making it perfect for deployment with the raspberry pi. Amazon Web Services (AWS) [22] was chosen as a cloud provider for storage of the data. The AWS provides services that make it easy to store images and also offers an initial free service for 1 year. Python programming language [23] was used in writing programs to capture the images using the Pi camera. The block diagram of the system for capturing the dataset is shown in Figure 1.

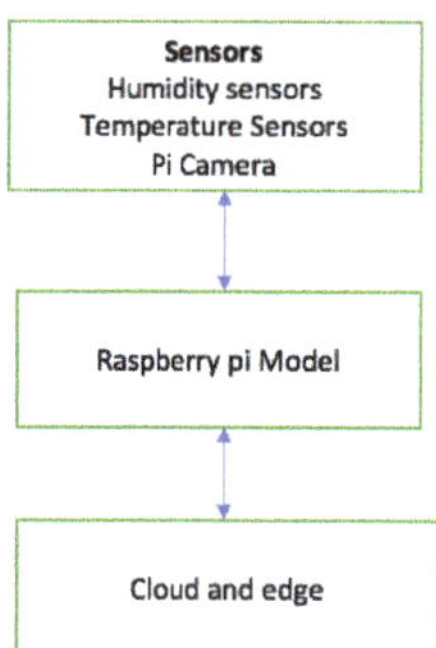

Figure 1. Block diagram of the data collection system.

2.2. Collection of the Dataset

The Internet-of-Things-based system for capturing the data was deployed in the Sisibo factory, Kenya for 4 days: 10–13 August 2020. The system was set up just above the tea fermentation bed (Figure 2). After collection of each image, the tea experts provided ground truths on their correct classes.

Figure 2. Collection of the dataset in Sisibo tea factory, Kenya using Raspberry pi and Pi camera.

3. Data Description

This study releases a black tea fermentation dataset which contains black tea fermentation images and physical parameters of tea during fermentation, which were collected in Sisibo tea factory, Kenya between 10 and 13 August 2020. The dataset can be found here: https://doi.org/10.5281/zenodo.4469326 (accessed on 2 March 2021). The file structure of the dataset is shown in Figure 3.

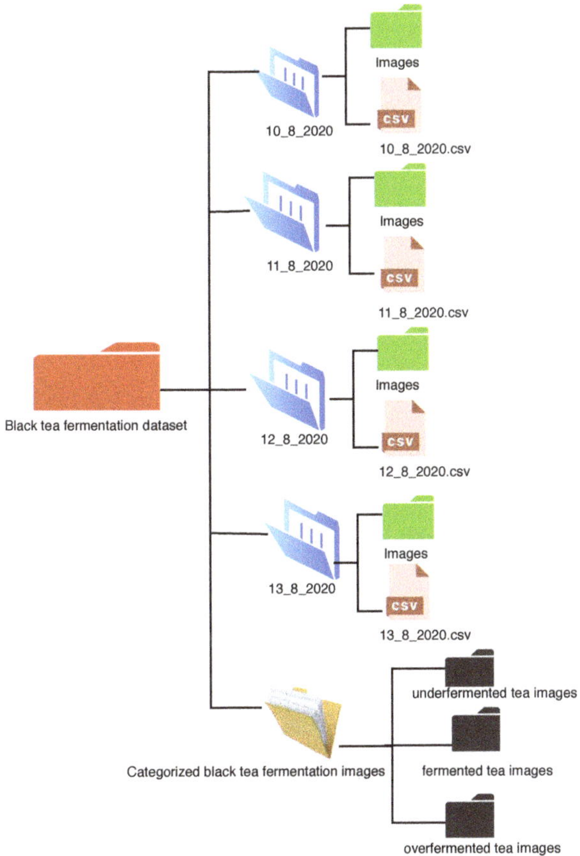

Figure 3. File structure of the black tea fermentation dataset.

Black tea fermentation conditions are contained in a Comma Delimited Values (CSV) file and contain the fermentation time, temperature, humidity, reference to image and the category of tea (Figure 4).

1	Time	Temperature (°C)	Humidity (%)	Reference to image	Category of tea
2	12:31:13	24	61.6	underfermented_1	unfermented
3	12:32:12	23.9	62.1	underfermented_2	unfermented
4	12:35:58	23.7	62.8	underfermented_3	unfermented
5	12:36:52	23.6	64	underfermented_4	unfermented
6	12:37:55	23.4	63.9	underfermented_5	unfermented
7	12:38:53	23.4	64.9	underfermented_6	unfermented
8	12:42:53	23.5	64.1	underfermented_7	unfermented
9	12:43:51	23.4	65	underfermented_8	unfermented
10	12:44:54	23.4	64.5	underfermented 9	unfermented

Figure 4. A Screenshot of the fermentation condition CSV file.

The images folder contains images that were taken as fermentation took place. These images correspond with the Reference showing the images column in the CSV files. In addition to these images, "Categorized black tea fermentation images" folder contains 6000 black tea fermentation images categorized into three classes of 2000 images each. The classification of these images was based on the decision of two tea fermentation experts. The experts relied on the color, smell, and taste of tea during fermentation to classify the images. The classes are unfermented, fermented, and overfermented.

The underfermented tea is in a folder labeled "unfermented tea" in the "Categorized black tea fermentation images" folder. The level of fermentation of this category of tea is below the optimum. These tea images are usually green. The images are labeled from "underfermented _00 " to "underfermented _1999". A sample of the images is presented in Figure 5a.

The fermented tea is optimally fermented and is usually coppery brown. The fermented tea is in a folder labeled "fermented tea" in the images folder. The images are labeled from "fermented _00" to "fermented _1999". A sample of the images is presented in Figure 5b.

The fermentation level of overfermented tea is beyond the optimum and is dark red. Overfermented tea is in a folder labeled "overfermeted tea". The images are labeled from "Overfermented _00" to "Overfermented _1999". A sample of the images is presented in Figure 5c.

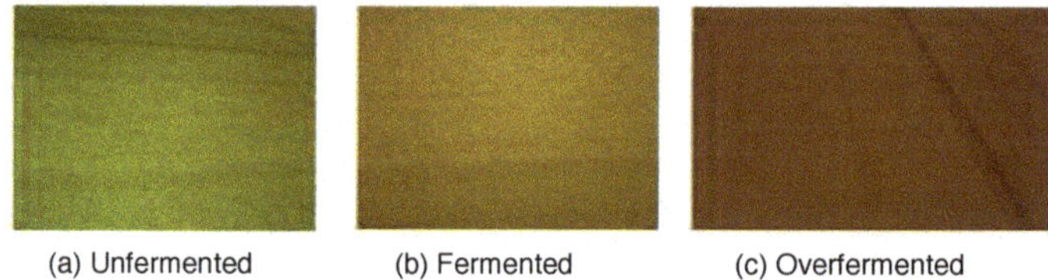

(a) Unfermented   (b) Fermented   (c) Overfermented

Figure 5. A sample of classes of tea image fermentation dataset released in this paper.

4. Data Validation

The fermentation of tea must take place within a given range of temperature and humidity. These ranges vary in different countries but, in Kenya, the acceptable range is between 20 and 30 degrees celcius. Figure 6 shows temperature and humidity values during a fermentation cycle in Sisibo tea factory on 10 August 2020. The fermentation process started at 12:31:13 h. The temperature recorded ranged from 19 °C to 28 °C , which was well within the optimum ranges. The values of the temperature increased steadily with time due to the natural climatic conditions of the area. On the other hand, humidity was between 75% and 92% for the low and the high, respectively. The values of humidity reduced steadily with time. The highest temperature value recorded was 28.8 °C, while the highest humidity value achieved was 82.6%. The fermentation curve was smooth and it took 68 min before the tea was fully fermented.

Figure 7 shows temperature and humidity values during a fermentation cycle in Sisibo tea factory on 11 August 2020. The fermentation process started at 08:48:03 h. The temperatures recorded ranged from 20.6 °C to 24.50 °C. These temperatures were within the optimum. The values of the temperature increased steadily with time due to the natural climatic conditions of the area. The highest humidity value recorded was at 90.8%. The values of humidity fluctuated throughout the fermentation period. The fermentation curve was smooth and it took 51 min before the tea was fully fermented. This is the shortest fermented duration reported in this data descriptor, as the fermentation temperatures were higher than the rest of the days.

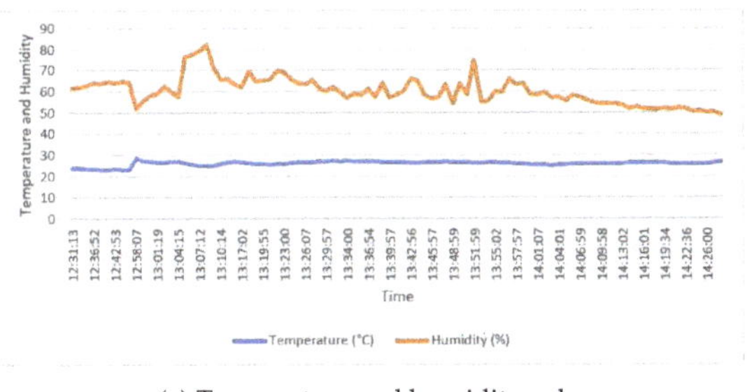

(**a**) Temperature and humidity values

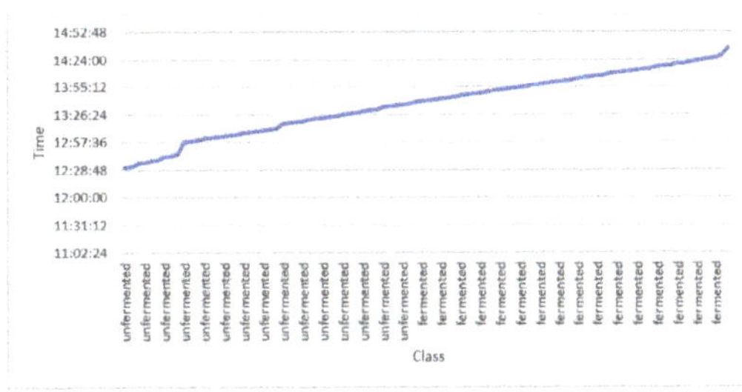

(**b**) Fermentation classes of tea with time

Figure 6. Fermentation conditions of tea in Sisibo tea factory on 10 August 2020.

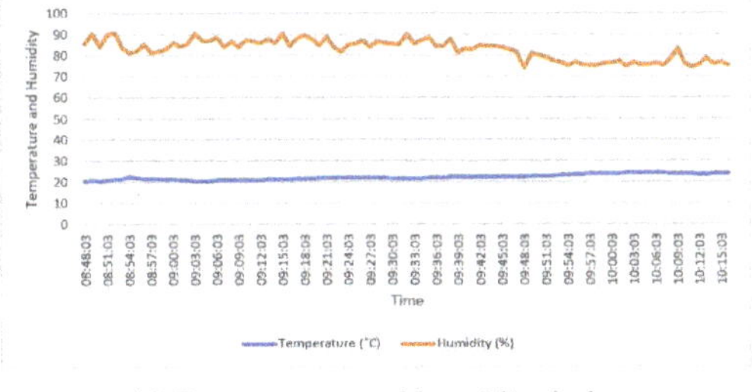

(**a**) Temperature and humidity values

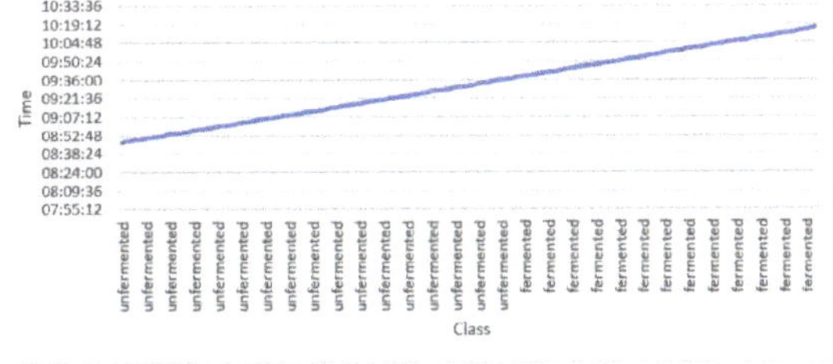

(**b**) Fermentation classes of tea with time

Figure 7. Fermentation conditions of tea in Sisibo tea factory on 11 August 2020.

Figure 8 shows temperature and humidity values during a fermentation cycle in Sisibo tea factory on 12 August 2020. The fermentation process started at 02:00:13 h. The temperature recorded ranged from 15 °C to 25.53 °C. These temperatures fell below the optimum, unlike on 10 August 2020. This is due to the fermentation occurring in the early morning when the region is naturally colder than during midday hours. The values of the temperature increased steadily with time, due to the natural climatic conditions of the area. The highest humidity value recorded was at 95%. The fermentation curve was smooth and it took 64 min before the tea was fully fermented.

Figure 9 shows temperature and humidity values during a fermentation cycle in Sisibo tea factory on 13 August 2020. The fermentation process started on 07:00:00 h. The highest temperature recorded was at 26.25 °C. Once again, the temperature ranges fell below the optimum, unlike on 10 August 2020. This is because fermentation was carried out in the early morning when the region is naturally colder than during midday hours. The highest humidity value recorded was at 95%. The fermentation curve was smooth and it took 58 min before the tea was fully fermented.

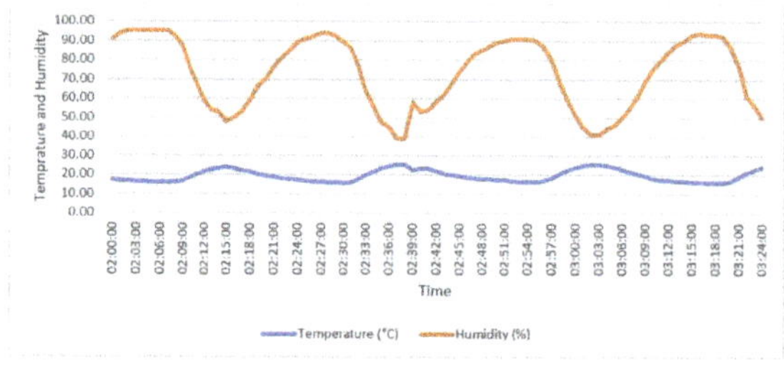

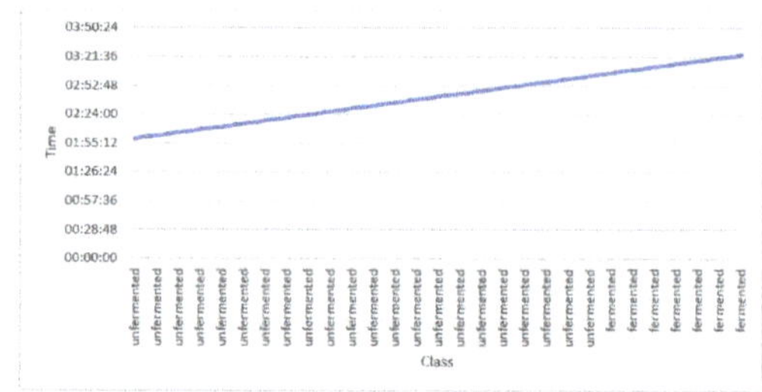

(**a**) Temperature and humidity values

(**b**) Fermentation classes of tea with time

Figure 8. Fermentation conditions of tea in Sisibo tea factory on 12 August 2020.

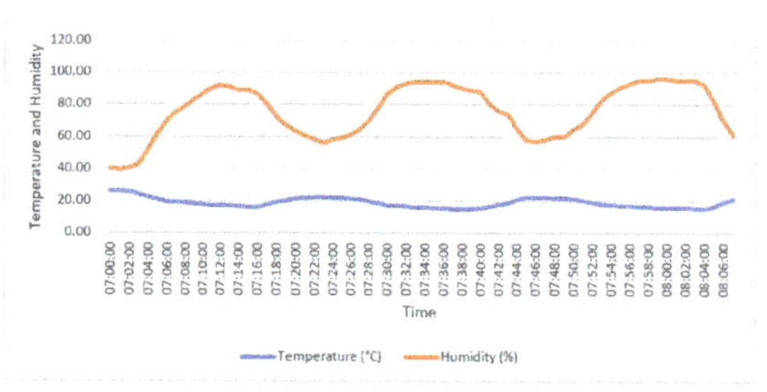

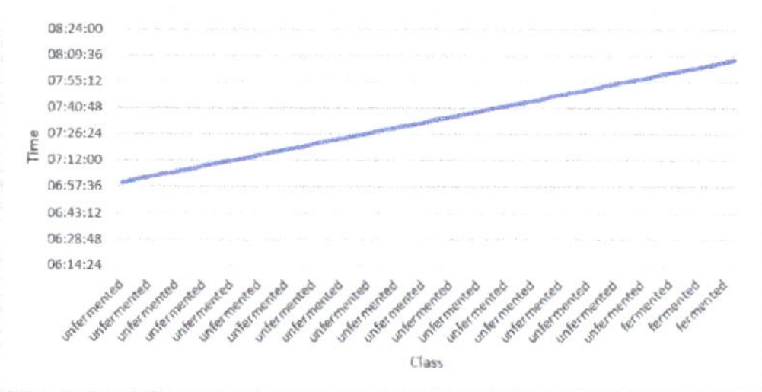

(**a**) Temperature and humidity values

(**b**) Fermentation classes of tea with time

Figure 9. Fermentation conditions of tea in Sisibo tea factory on 13 August 2020.

The average temperature and humidity values for the days are presented in Figure 10. The highest average temperature was recorded on 10 August 2020, with 26.14 °C. On 11 August 2020, an average temperature of 22.31 °C was recorded. On the other days, the average temperatures were 19.39 °C and 18.83 °C for 12 August 2020 and 3 August 2020, respectively. Thus, the fermentation took place outside the range of 20–30 °C for the two days.

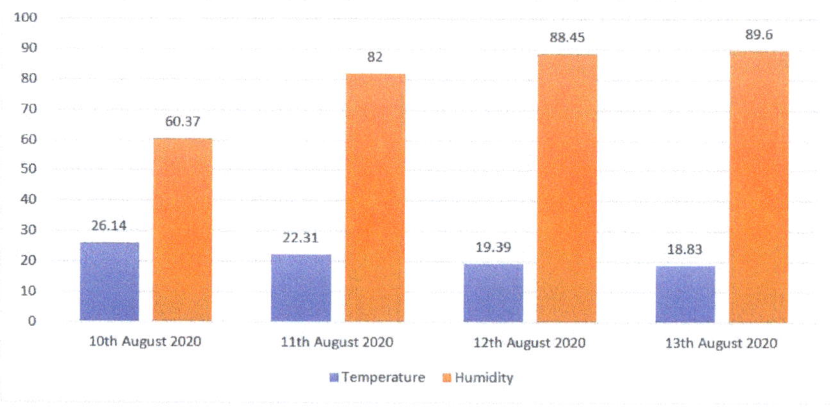

Figure 10. Temperature and humidity during black tea fermentation in Sisibo tea factory between 10 and 13 August 2020.

5. Conclusions

In this paper, a tea fermentation dataset has been released. The dataset was collected in the Sisibo tea factory, Kenya using an IoT-based model. The model predicted fermentation cycle of tea and correlated well with the decisions of the tea fermentation experts. Fermentation experts gave ground truths for the dataset. The dataset can be used by researchers in training machine learning models for the detection of the optimum fermentation of tea. This is a significant achievement in the field of applying machine learning to the detection of optimum fermentation of tea, as there is currently only one existing dataset for training these models.

Author Contributions: Formal analysis, A.F.; methodology, G.K. and A.F.; software, G.K.; supervision, A.N., R.N.S., R.C.R., and A.F.; validation, G.K.; visualization, G.K.; writing original draft, G.K.; writing review and editing, A.N., R.N.S., R.C.R., and A.F. All authors have read and agreed to the published version of the manuscript.

Funding: The Article Processing Charges was funded by the University of Bremen, Germany. This research received financial support from the African Center of Excellence in Internet of Things (ACEIoT) and the Kenya Education Network (KENET) through the Computer Science, Information Systems, and Engineering multidisciplinary mini-grant.

Institutional Review Board Statement: Not applicable.

Informed Consent Statement: Not applicable.

Data Availability Statement: The data presented in this study are publicly archived in the Zenodo platform: https://doi.org/10.5281/zenodo.4469326 (accessed on 2 March 2021).

Acknowledgments: We wish to thank the University of Bremen for funding the Article Processing Charges of this article. Additionally, we wish to thank the management and the entire staff of Sisibo tea factory, Kenya for allowing us to collect the datasets in their factory. Special appreciation goes to the tea tasters of Sisibo Tea Factory: Abdalla Tunya and Cornellius Chirchir, who offered their expert judgement for every tea image in the dataset. The authors wish to acknowledge the Kenya Education Network (KENET) for financially supporting this research through the CS/IT Minigrant. Finally, the authors would like to express their appreciation to the anonymous reviewers and editors for their constructive comments.

Conflicts of Interest: The authors declare no conflict of interest.

Abbreviations

The following abbreviations are used in this manuscript:

IoT	Internet of Things
AWS	Amazon Web Services
E-Commerce	Electronic Commerce

References

1. Miura, K.; Hughes, M.C.B.; Arovah, N.I.; Van Der Pols, J.C.; Green, A.C. Black Tea Consumption and Risk of Skin Cancer: An 11-Year Prospective Study. *Nutr. Cancer* **2015**, *67*, 1049–1055. [CrossRef] [PubMed]
2. Saikia, D.; Boruah, P.K.; Sarma, U. A Sensor Network to Monitor Process Parameters of Fermentation and Drying in Black Tea Production. *Mapan* **2015**, *30*, 211–219. [CrossRef]
3. Bhattacharyya, N.; Seth, S.; Tudu, B.; Tamuly, P.; Jana, A.; Ghosh, D.; Bandyopadhyay, R.; Bhuyan, M.; Sabhapandit, S. Detection of optimum fermentation time for black tea manufacturing using electronic nose. *Sens. Actuators B Chem.* **2007**, *122*, 627–634. [CrossRef]
4. Kimutai, G.; Ngenzi, A.; Said, R.N.; Kiprop, A.; Förster, A. An Optimum Tea Fermentation Detection Model Based on Deep Convolutional Neural Networks. *Data* **2020**, *5*, 44. [CrossRef]
5. Onduru, D.D.; De Jager, A.; Hiller, S.; Van Den Bosch, R. Sustainability of Smallholder Tea Production in Developing Countries: Learning Experiences from Farmer Field Schools in Kenya. *Int. J. Dev. Sustain.* **2012**, *1*, 714–742.
6. Tea Board of Kenya. *Kenya Tea Yearly Report*; Technical Report; Tea Board of Kenya: Nairobi, Kenya, 2018.
7. Kagira, E.K.; Kimani, S.W.; Githii, K.S. Sustainable Methods of Addressing Challenges Facing Small Holder Tea Sector in Kenya: A Supply Chain Management Approach. *J. Manag. Sustain.* **2012**, *2*. [CrossRef]
8. Kamunya, S.M.; Wachira, F.N.; Pathak, R.S.; Muoki, R.C.; Sharma, R.K. Tea Improvement in Kenya. In *Advanced Topics in Science and Technology in China*; Springer: Berlin/Heidelberg, Germany, 2012; pp. 177–226. [CrossRef]
9. Ghosh, S.; Tudu, B.; Bhattacharyya, N.; Bandyopadhyay, R. A recurrent Elman network in conjunction with an electronic nose for fast prediction of optimum fermentation time of black tea. *Neural Comput. Appl.* **2019**, *31*, 1165–1171. [CrossRef]

10. Deb, S.; Jolvis Pou, K.R. A Review of Withering in the Processing of Black Tea. *J. Biosyst. Eng.* **2016**, *41*, 12. [CrossRef]
11. Binh, P.T.; Du, D.H.; Nhung, T.C. Control and Optimize Black Tea Fermentation Using Computer Vision and Optimal Control Algorithm. In *Lecture Notes in Networks and Systems*; Springer: Berlin/Heidelberg, Germany, 2020; Volume 104, pp. 310–319. [CrossRef]
12. Ghosh, A.; Sharma, P.; Tudu, B.; Sabhapondit, S.; Baruah, B.D.; Tamuly, P.; Bhattacharyya, N.; Bandyopadhyay, R. Detection of Optimum Fermentation Time of Black CTC Tea Using a Voltammetric Electronic Tongue. *IEEE Trans. Instrum. Meas.* **2015**, *64*, 2720–2729. [CrossRef]
13. Debashis, S.; Boruah, P.K.R.; Sarma, U. Development and implementation of a sensor network to monitor fermentation process parameter in tea processing. *Int. J. Smart Sens. Intell. Syst.* **2014**, *7*, 1254–1270.
14. Jolvis Pou, K. Fermentation: The Key Step in the Processing of Black Tea. *J. Biosyst. Eng.* **2016**, *41*, 85–92. [CrossRef]
15. Manigandan, N. Handheld Electronic Nose (HEN) for detection of optimum fermentation time during tea manufacture and assessment of tea quality. *Int. J. Adv. Res.* **2019**, *7*, 697–702. [CrossRef]
16. Shen, L.; Margolies, L.R.; Rothstein, J.H.; Fluder, E.; McBride, R.; Sieh, W. Deep Learning to Improve Breast Cancer Detection on Screening Mammography. *Sci. Rep.* **2019**, *9*. [CrossRef] [PubMed]
17. Kimutai, G.; Cheruiyot, P.W.; Otieno, D.C. A Content Based Image Retrieval Model for E-Commerce. *Int. J. Eng. Comput. Sci.* **2018**, *7*, 24392–24396. [CrossRef]
18. Chityala, R.; Pudipeddi, S. *Image Processing and Acquisition using Python*, 1st ed.; Chapman and Hall/CRC: Boca Raton, FL, USA, 2014; p. 390.
19. Kimutai, G.; Anna, F. *Black Tea Fermentation Dataset*; Technical Report; Mendeley Ltd.: London, UK, 2020. [CrossRef]
20. Marot, J.; Bourennane, S. Raspberry Pi for image processing education. In Proceedings of the 25th European Signal Processing Conference, EUSIPCO 2017, Kos, Greece, 28 August–2 September 2017; Volume 2017, pp. 2364–2368. [CrossRef]
21. Thangavel, S.K.; Murthi, M. A semi automated system for smart harvesting of tea leaves. In Proceedings of the 2017 4th International Conference on Advanced Computing and Communication Systems, ICACCS 2017, Coimbatore, India, 6–7 January 2017. [CrossRef]
22. Narula, S.; Jain, A.; Prachi. Cloud computing security: Amazon web service. In Proceedings of the International Conference on Advanced Computing and Communication Technologies, ACCT, Haryana, India, 21–22 February 2015; Volume 2015, pp. 501–505. [CrossRef]
23. Dubosson, F.; Bromuri, S.; Schumacher, M. A python framework for exhaustive machine learning algorithms and features evaluations. In Proceedings of the International Conference on Advanced Information Networking and Applications, AINA, Crans-Montana, Switzerland, 23–25 March 2016; Volume 2016, pp. 987–993. [CrossRef]

 *data*

Data Descriptor

A Probabilistic Bag-to-Class Approach to Multiple-Instance Learning

Kajsa Møllersen [1,*] , Jon Yngve Hardeberg [2] and Fred Godtliebsen [3]

[1] Department of Community Medicine, Faculty of Health Science, UiT The Arctic University of Norway, N-9037 Tromsø, Norway
[2] Department of Computer Science, Faculty of Information Technology and Electrical Engineering, NTNU—Norwegian University of Science and Technology, N-2815 Gjøvik, Norway; jon.hardeberg@ntnu.no
[3] Department of Mathematics and Statistics, Faculty of Science and Technology, UiT The Arctic University of Norway, N-9037 Tromsø, Norway; fred.godtliebsen@uit.no
* Correspondence: kajsa.mollersen@uit.no; Tel.: +47-9778-3940

Received: 31 March 2020; Accepted: 23 June 2020; Published: 26 June 2020

Abstract: Multi-instance (MI) learning is a branch of machine learning, where each object (bag) consists of multiple feature vectors (instances)—for example, an image consisting of multiple patches and their corresponding feature vectors. In MI classification, each bag in the training set has a class label, but the instances are unlabeled. The instances are most commonly regarded as a set of points in a multi-dimensional space. Alternatively, instances are viewed as realizations of random vectors with corresponding probability distribution, where the bag is the distribution, not the realizations. By introducing the probability distribution space to bag-level classification problems, dissimilarities between probability distributions (divergences) can be applied. The bag-to-bag Kullback–Leibler information is asymptotically the best classifier, but the typical sparseness of MI training sets is an obstacle. We introduce bag-to-class divergence to MI learning, emphasizing the hierarchical nature of the random vectors that makes bags from the same class different. We propose two properties for bag-to-class divergences, and an additional property for sparse training sets, and propose a dissimilarity measure that fulfils them. Its performance is demonstrated on synthetic and real data. The probability distribution space is valid for MI learning, both for the theoretical analysis and applications.

Dataset: Breast tissue images available at https://bioimage.ucsb.edu/research/bio-segmentation, extracted feature vectors available at https://figshare.com/articles/MIProblems_A_repository_of_multiple_instance_learning_datasets/6633983. *BreakHis* data available at https://web.inf.ufpr.br/vri/databases/breast-cancer-histopathological-database-breakhis/. Code available at https://github.com/kajsam/ProbabilisticBag2Class.

Dataset License: CC BY 4.0

Keywords: image classification; multi-instance learning; divergence; dissimilarity; bag-to-class; Kullback–Leibler

---

1. Introduction

1.1. Classification of Weakly Supervised Data

Machine-learning applications include a wide variety of data types, images being one of the most successful areas. It has had an enormous impact on image analysis, especially in replacing small sets of hand-crafted features with large sets of computer readable features, which often lack apparent or intuitive meaning. The task and problems to which machine learning is applied can be divided broadly into unsupervised and supervised learning. In supervised learning, the training data consists of K objects, $\mathbf{x}$, with corresponding class labels, y; $\{(\mathbf{x}_1, y_1), \ldots, (\mathbf{x}_k, y_k), \ldots, (\mathbf{x}_K, y_K)\}$. An object is typically a vector of d feature values, $\mathbf{x}_k = (x_{k1}, \ldots, x_{kd})$, observed directly or extracted from e.g., an image. In classification, the task is to build a classifier that correctly labels a new object. The training data is used to adjust the model according to the desired outcome, often maximizing the accuracy of the classifier.

For many types of images, only a small part of the image defines the class, but the label is available only at image level. This is common in medical images, such as histology slides, where the tumor cells typically make up a small proportion of the image. However, the location of those cells is not available for training. Multi-instance (MI) learning is a branch of machine learning that specifically targets problems where labels are available only at a superior level, and relates to other weakly supervised data problems, such as semi-supervised learning and transfer learning through label scarcity [1].

1.2. Multi-Instance Learning

In MI learning, each object is a set of feature vectors referred to as instances. The set $\mathbb{X}_k = \{\mathbf{x}_{k1}, \ldots, \mathbf{x}_{kn_k}\}$, where the n_k elements are vectors of length d, is referred to as bag. The number of instances, n_k, varies from bag to bag, whereas the vector length is constant. In supervised MI learning, the training data consists of K sets and their corresponding class labels, $\{(\mathbb{X}_1, y_1), \ldots, (\mathbb{X}_k, y_k), \ldots, (\mathbb{X}_K, y_K)\}$.

Figure 1a shows an image (bag), k, of benign breast tissue [2], divided into n_k segments with corresponding feature vectors (instances) $\mathbf{x}_{k1}, \ldots, \mathbf{x}_{kn_k}$ [3]. Correspondingly, Figure 1b shows malignant breast tissue.

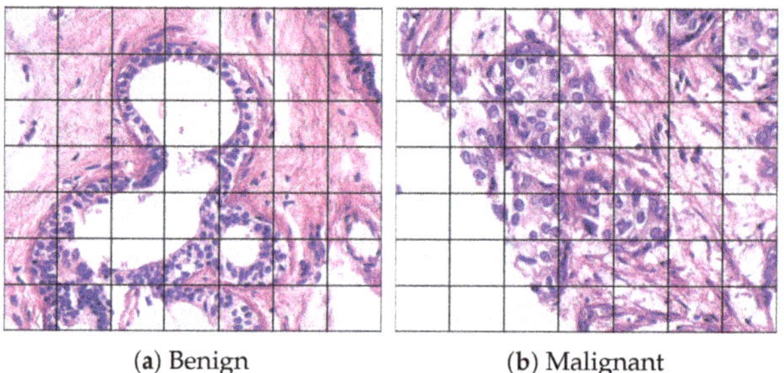

(a) Benign   (b) Malignant

Figure 1. Breast tissue images [2]. The image segments are not labeled.

The images in the data set have class labels; however, the individual segments do not. This is a key characteristic of MI learning—the instances are not labeled. MI learning includes instance classification [4], clustering [5], regression [5], and multi-label learning [6,7], but this article will focus on bag classification. MI learning can also be found as integrated parts of end-to-end methods for image analysis that generate patches, extract features and do feature selection [7]. See also [8] for an overview and discussion on end-to-end neural network MI learning methods.

The term "MI learning" was introduced in an application of molecules (bags) with different shapes (instances), and their ability to bind to other molecules [9]. A molecule binds if at least one of its shapes can bind. In MI terminology, the classes in binary classification are referred to as positive, *pos*, and negative, *neg*. The assumption that a positive bag contains at least one positive instance, and a negative bag contains only negative instances is referred to as the standard MI assumption.

Many new applications violate the standard MI assumption, such as image classification [10] and text categorization [11]. Consequently, successful algorithms meet more general assumptions, see e.g., the hierarchy of Weidmann et al. [12] or Foulds and Frank's taxonomy [13]. For a more recent review of MI classification algorithms, see e.g., [14]. Amores [15] presented the three paradigms of instance space (IS), embedded space (ES), and bag space (BS). IS methods aggregate the outcome of single-instance classifiers applied to the instances of a bag, whereas ES methods map the instances to a vector, followed by use of a single-instance classifier. In the BS paradigm, the instances are transformed to a non-vectorial space where the classification is performed, avoiding the detour via single-instance classifiers. The non-vectorial space of probability functions has not yet been introduced to the BS paradigm, despite its analytical benefits, see Sections 3.2 and 3.3.

Although both Carbonneau et al. [16] and Amores [15] defined a bag as a set of feature vectors, Foulds and Frank [13] stated that a bag can also be modelled as a probability distribution. The distinction is necessary in analysis of classification approaches, and both viewpoints offer benefits, see Section 6.1 for a discussion.

1.3. Bag Density and Class Sparsity

Optimal classification in MI learning depends on the number of instances per bag (bag density) and the number of bags per class in the training set (class density). Sample sparsity is a common obstacle in MI learning [16], which we address in Section 3.5. High bag density ensures a precise description of each bag, whereas high class density ensures precise modelling of each class when training the classifier. In image analysis, the number of patches corresponds to the number of instances, and is commonly a user input parameter. The number of images corresponds to the number of bags, and is limited by the training set itself.

High resolution of today's images and the increasingly common practice of sharing the images themselves instead of extracted features ensure high bag density. The number of bags available for training is still limited, and will continue to be so in the foreseeable future, especially for medical images where data collection is restricted by laws and regulations. This motivates an approach to MI learning that can exploit the increasing bag density and overcome the class sparsity.

1.4. A Probabilistic Bag-to-Class Approach to Multi-Instance Learning

We propose to model the bags as probability distributions and the instances as random samples. The bags are assumed to be random samples from their respective classes and the instance-bag sampling form a hierarchical distribution. Hierarchical distribution is novel for bag classification and novel outside the strict standard MI assumption. Unbiased estimators for the bag probability distributions ensure that as the number of instances increases ($n_k \rightarrow \infty$), the discrepancy between the estimate and the underlying truth diminishes, taking advantage of increasing bag density. To overcome the problem of class sparsity, the instances are aggregated at class level.

We further propose to use a bag-to-class dissimilarity measure for classification. This is novel in the MI context, where dissimilarity measures have been either instance-to-instance or bag-to-bag. With the analytical framework of probability distributions and their dissimilarity measures, we present the optimal classifier for dense class sampling as a theoretical background and identify data-independent properties for bag classification under class sparsity.

The main contribution of this article is a bag-to-class dissimilarity measure for sparse training data. It builds on:

- presenting the hierarchical model for general, non-standard MI assumptions (Section 3.3),
- introduction of bag-to-class dissimilarity measures (Section 3.5), and
- identification of two properties for bag-to-class divergence (Section 4.1).

The novelty is that it takes into account the class sparsity by comparing a bag to one class while conditioning on the other class.

In Section 5, the Kullback–Leibler (KL) information and the new dissimilarity measure is applied to data sets and the results are reported. Bags defined in the probability distribution space, in combination with bag-to-class divergence, constitutes a new framework for MI learning, which is compared to other frameworks in Section 6.

2. Related Work and State-of-the-Art

The feature vector set viewpoint seems to be the most common, but the probabilistic viewpoint was introduced already in 1998, under the assumption that instances of the same class are independent and identically distributed (i.i.d.) [17]. This assumption has been used in approaches such as estimating the expectation by the mean [18], or estimation of class distribution parameters [19], but has also been criticized [20]. The hierarchical distribution was introduced for learnability theory under the standard MI assumption for instance classification in 2016 [4]. We expand the use for more general assumptions in Section 3.3.

Dissimilarities in MI learning have been categorized as instance-to-instance or bag-to-bag [15,21]. The bag-to-prototype approach in [21] offers an in-between category, but the theoretical framework is missing. Bag-to-class dissimilarity has not been studied within the MI framework, but has been used under the i.i.d. given class assumption for image classification [22]. The sparseness of training sets was also addressed: if the instances are aggregated on class level, a denser representation is achieved. Many MI algorithms use dissimilarities, e.g., graph distances [23], Hausdorff metrics [24], functions of the Euclidean distance [14,25], and distribution parameter-based distances [14]. The performances of dissimilarities on specific data sets have been investigated [14,19,21,25,26], but more analytical comparisons are missing. A large class of commonly used kernels are also distances [27], and hence, many kernel-based approaches in MI learning can be viewed as dissimilarity-based approaches. In [28], the Fisher kernel is used as input to a support vector machine (SVM), whereas in [11,20] the kernels are an integrated part of the methods.

The non-vectorial graph space was used in [20,23]. We introduce the non-vectorial space of probability functions as an extension within the BS paradigm for bag classification through dissimilarity measures between distributions in Section 3.2.

The KL information was applied in [22], and is a much-used divergence function. It is closely connected to the Fisher information [29] used in [28] and to the cross entropy used as loss function in [8]. We propose a conditional KL information in Section 4.2, which differs from the earlier proposed weighted KL information [30] whose weight is a constant function of X.

There is a wide variety in MI learning, both in methods and data sets, and it should be clear that state-of-the-art will depend on the type of data. Sudharshan et al. [31] gave a comparison of 12 MI classification methods and five state-of-the-art general classification methods on a well-described, publicly available histology image data set. All methods included have shown best performance on other data sets. The five methods that showed best performance for at least one of the data subsets serve as state-of-the-art baseline for evaluation in Section 5.3.

Cheplygina et al. [1] gave an overview of MI learning applications in different categories, but no comparison was made. The work of Sudharshan et al. falls into the "Histology/Microscopy" category, and the overview offers a potential expansion of histology state-of-the-art. Among the 12 listed articles, Zhang et al. [32] concluded that GPMIL outperforms Citation-kNN, which is one of the 12 methods in [31], but not one of the 5 best-performing. Kandemir et al. [3], Li et al. [33] and Tomczak et al. [34] presented methods that outperform GPMIL on a publicly available data set. We include these as comparison.

Of the remaining articles, none of them present an extensive comparison to other methods, their data sets are either non-public [35–38], no longer available [39], or the reference is not complete [40,41], which make them unsuitable for comparison. Jia et al. [42] presented a segmentation method, and is therefore not comparable.

3. Theoretical Background and Intuitions

3.1. Notation

Subscript and superscript *pos* and *neg* refer to the class label of the bag, subscript and superscript $+$ and $-$ refer to the unknown instance label.

X : instance random vector
C : class, either *pos* or *neg*
B : bag
$P(\cdot)$: probability distribution
x_{ki} : feature vector (instance) in set k, $i = 1, \ldots, n_k$
$\mathbb{X}_k$: set of feature vectors k of size n_k
y_k: bag label
$\mathcal{X}$: sample space for instances
$\mathcal{X}^+$: sample space for positive instances
$\mathcal{X}^-$: sample space for negative instances
$\mathcal{B}_{pos}$: sample space of positive bags
$\mathcal{B}_{neg}$: sample space of negative bags
$P(C|\mathbb{X}_k)$: posterior class probability, given instance sample
Θ : parameter random vector
θ_k : parameter vector
$P_{bag}(X) = P(X|B)$: probability distribution for instances in bag B
$P(X|\theta_k)$: parameterized probability distribution of bag k
$P_{pos}(X) = P(X|pos)$: probability distribution for instances from the positive class
$P_{neg}(X) = P(X|neg)$: probability distribution for instances from the negative class
τ_i : instance label
π_k : probability of positive instances
$D(P_k, P_\ell) = D(P_k(X), P_\ell(X))$: divergence from $P_k(X)$ to $P_\ell(X)$
$f_k(\mathbf{x}) = f(\mathbf{x}|\theta_k)$: probability density function (PDF) for bag k
$D(f_k, f_\ell) = D(f_k(\mathbf{x}), f_\ell(\mathbf{x}))$: divergence from $f_k(\mathbf{x})$ to $f_\ell(\mathbf{x})$

We assume $P(X) < \infty$, and equivalently $f(\mathbf{x}) < \infty$, for all distributions.

3.2. The Non-Vectorial Space of Probability Functions

The intuition behind the probabilistic approach in MI learning can be understood through image analysis and tumor classification. Figure 1a represents parts of a tumor, chosen carefully for diagnostic purposes. The process from biological material to image contains steps whose outcome is influenced by subjective choices and randomness: The precise day the patient is admitted influences the state of the tumor; the specific parts of the tumor that are extracted for staining; the actual stain varies from batch to batch, and the imaging equipment has multiple parameter settings. All this means that the same tumor would have produced a different image under different circumstances. The process from image to feature vector set also contains several steps: Patch size, grid or random patches, color conversion, etc. In summary, the observed feature vectors are a representation of an underlying object, and that representation may vary, even if the object remains fixed.

From the probabilistic viewpoint, an instance, $\mathbf{x}$, is a realization of a random vector, X, with probability distribution $P(X)$ and sample space $\mathcal{X}$. The bag is the probability distribution $P(X)$, and the set of instances, $\mathbb{X}$, is multiple realizations of X. The task of an MI classifier is to classify the bag given the observations, $\mathbb{X}$.

The posterior class probability, $P(C|\mathbb{X}_k)$, is an effective classifier if the standard MI assumption holds, since it is defined as:

$$P(pos|\mathbb{X}_k) = \begin{cases} 1 \text{ if any } \mathbf{x}_{ki} \in \mathcal{X}^+, i = 1, \ldots, n_k \\ 0 \text{ otherwise,} \end{cases}$$

where $\mathcal{X}^+$ is the positive instance space, and the positive and negative instance spaces are disjoint.

Bayes' rule, $P(C|X) \propto P(X|C)P(C)$, can be used when the posterior probability is unknown. An assumption used to estimate the probability distribution of instance given the class, $P(X|C)$, is that instances from bags of the same class are i.i.d. random samples. However, this is a poor description for MI learning.

3.3. Hierarchical Distributions

As an illustrative example, let the instances be the color of image patches from the class *sea* or *desert*, and let image k depict a blue sea like in Figure 2a with instances $\mathbb{X}_k$, and image ℓ depict a turquoise sea like in Figure 2b with instances $\mathbb{X}_\ell$. The instances are realizations from $P(X|\theta_k)$ and $P(X|\theta_\ell)$, respectively, where θ is the parameter indicating the colors. If the instance distribution were dependent only on class, then $\theta_k = \theta_\ell = \theta_{sea}$, which is clearly not the case. Instance distributions are dependent not only on class, but also on bag. The random vectors in $\mathbb{X}_k$ are i.i.d., but have a different distribution than those in $\mathbb{X}_\ell$. An important distinction between uncertain objects, whose distribution depends solely on the class label [43,44], and MI learning is that the instances of two bags from the same class are not from the same distribution.

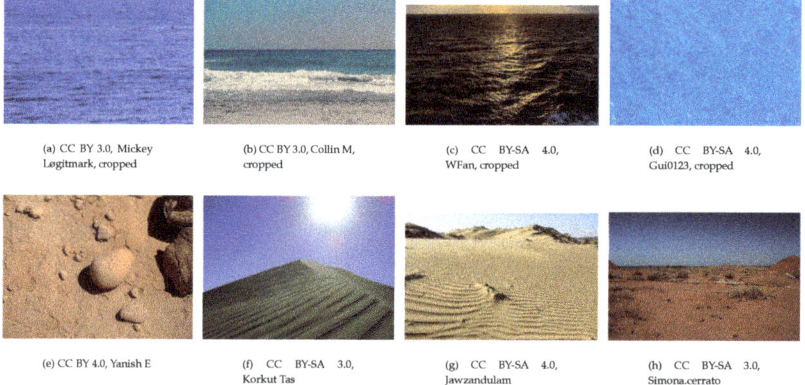

Figure 2. Sea and desert images from Wikimedia Commons.

The dependency nature for MI learning can be described as a hierarchical distribution (Equation (1)), where a bag, B, is defined as the probability distribution of its instances, $P(X|B)$, and the bag space, $\mathcal{B}$, is a set of distributions. A bag is itself a realization from the sample space of bags, whose distribution depends on the class. The generative model of instances from a positive or negative bag follows a hierarchical distribution:

$$X|B \sim P(X|B) \qquad X|B \sim P(X|B)$$
$$B \sim P(B|pos) \quad \text{or} \quad B \sim P(B|neg), \tag{1}$$

respectively. From a practical viewpoint, $P(X|B)$ can be considered parametric functions, $P(X|\theta)$, where the sampling of a bag corresponds to sampling the parameter vector θ that defines its distribution:

$$X|\theta_{pos} \sim P(X|\theta_{pos}) \qquad X|\theta_{neg} \sim P(X|\theta_{neg})$$
$$\Theta_{pos} \sim P(\Theta_{pos}) \quad \text{or} \quad \Theta_{neg} \sim P(\Theta_{neg}). \tag{2}$$

The parametric generative model is shown in Figure 3.

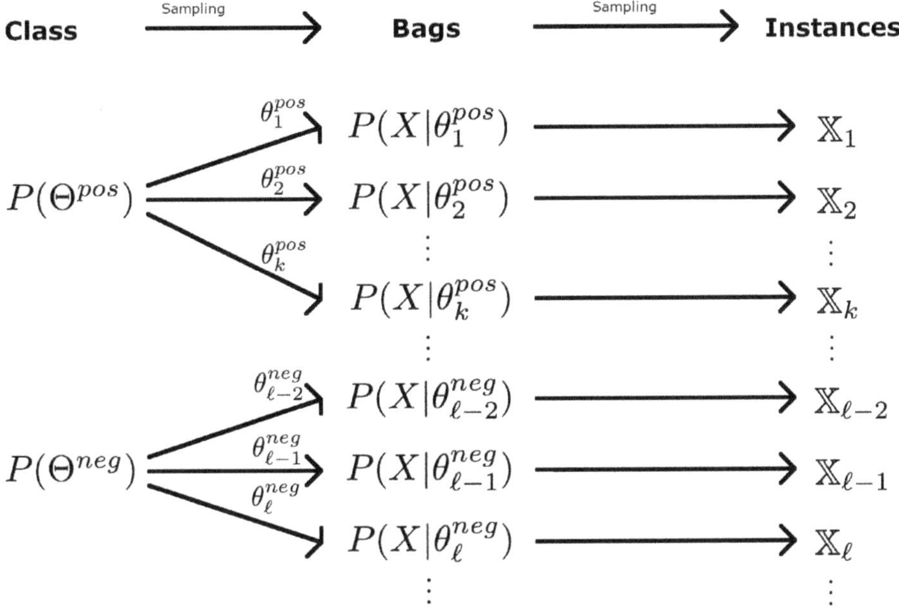

Figure 3. Parametric generative model. Bags are realizations of random parameter vectors, sampled according to the respective class distributions. Instances are realizations of feature vectors, sampled according the respective bag distributions. Only the instance sets are observed.

The common view in MI learning is that a bag consists of positive and negative instances, which corresponds to a bag being a mixture of a positive and a negative distribution. Consider tumor images labeled *pos* or *neg*, with instances extracted from patches. Let $P(X|\theta_k^+)$ and $P(X|\theta_k^-)$ denote the probability distributions of positive and negative segments, respectively, of image k. The distribution of bag k is a mixture distribution:

$$P(X|\pi_k, \theta_k^+, \theta_k^-) = p_k P(X|\theta_k^+) + (1 - p_k) P(X|\theta_k^-),$$

where $p_k = \sum_{i=1}^{n_k} \tau_i / n_k$, where $\tau_i = 1$ if instance i is positive. The parameter vector $(\pi_k, \theta_k^+, \theta_k^-)$ defines the bag. The probability of positive segments, π_k, depends on the image's class label, and hence π_k is sampled from $P(\Pi_{pos})$ or $P(\Pi_{neg})$. The characteristics of positive and negative segments vary from image to image. Hence, θ_k^+ and θ_k^- are realizations of random variables, with corresponding probability distributions $P(\Theta^+)$ and $P(\Theta^-)$. The generative model of instances from a positive bag is:

$$X|\tau, \theta^+, \theta^- \sim \begin{cases} P(X|\tau = 1) = P(X|\theta^+) \\ P(X|\tau = 0) = P(X|\theta^-) \end{cases}$$

$$\mathcal{T}|\pi_{pos} \sim \begin{cases} P(\tau = 1) = \pi_{pos} \\ P(\tau = 0) = 1 - \pi_{pos} \end{cases} \tag{3}$$

$$\Pi_{pos} \sim P(\Pi_{pos}), \quad \Theta^+ \sim P(\Theta^+), \quad \Theta^- \sim P(\Theta^-).$$

The corresponding sampling procedure from positive bag, k, is

Step 1: Draw π_k from $P(\Pi_{pos})$, θ_k^+ from $P(\Theta^+)$, and θ_k^- from $P(\Theta^-)$. These three parameters define the bag.
Step 2: For $i = 1, \ldots, n_k$, draw τ_i from $P(\mathcal{T}|\pi_k)$, draw x_i from $P(X|\theta_k^+)$ if $\tau_i = 1$, and from $P(X|\theta_k^-)$ otherwise.

The generative model and sampling procedure for negative bags are equivalent to that of positive bags.

By imposing restrictions, assumptions can be accurately described, e.g., the standard MI assumption: at least one positive instance in a positive bag: $P(p_k \geq 1/n_k) = 1$; no positive instances in a negative bag: $P(\Pi_{neg} = 0) = 1$; the positive and negative instance spaces are disjoint.

Equation (3) is the generative model of MI problems, assuming that the instances have unknown class labels and that the distributions are parametric. The parameters π_k, θ_k^+ and θ_k^- are i.i.d. samples from their respective distributions, but are not observed and are hard to estimate due to the very nature of MI learning: the instances are not labeled. Instead, $P(X|B)$ can be estimated from the observed instances, and a divergence function can serve as classifier.

The instance i.i.d. assumption is not inherent to the probability distribution viewpoint, but the asymptotic results for the KL information discussed in Section 3.5 rely on it. In many applications, such as image analysis with sliding windows, the instances are best represented as dependent samples, but the dependencies are hard to estimate, and the independence assumption is often the best approximation. Doran and Ray [4] showed that the independence assumption is an approximation of dependent instances, but comes with the cost of slower convergence.

3.4. Dissimilarities in MI Learning

The information contained at bag-level is converted before it is fed into a classifier. If the bags are sets, they are commonly converted into distances. Dissimilarities in MI learning can be categorized as instance-to-instance, bag-to-bag or bag-to-class. Amores [15] implicitly assumed metricity for dissimilarity functions [27] in the BS paradigm, but there is nothing inherent to MI learning that imposes these restrictions. In the case where bags are probability distributions, distances are no longer applicable since they live in a non-vectorial space. Distances are a special case of dissimilarity functions, and the equivalent for probability distributions are referred to as divergences, $D(P_k(X), P_\ell(X))$. Although distances fulfil three properties by definition—among them symmetry and zero distance for identical sets—divergences do not have such properties, by definition.

A group of divergences named f-divergences has properties that are reasonable to demand for measuring the ability to distinguish probability distributions [45,46]:

Equality and orthogonality: An f-divergence takes its minimum when the two probability functions are equal and its maximum when they are orthogonal. This means that two identical bags will have minimum dissimilarity between them, and that two bags without shared sample space will have maximum dissimilarity. A definition of orthogonal distributions can be found in [47].

Monotonicity: The f-divergences possess a monotonicity property that can be thought of as an equivalent to the triangle property for distances: For a family of densities with monotone likelihood ratio, if $a < \theta_1 < \theta_2 < \theta_3 < b$, then $D(P(X|\theta_1), P(X|\theta_2)) \leq D(P(X|\theta_1), P(X|\theta_3))$. This is valid, e.g., for Gaussian densities with equal variance and mean θ. This means that if the distance between θ_1 and θ_3 is larger than the distance between θ_1 and θ_2, the divergence is larger or equal. The f-divergences are not symmetric by definition, but some of them are.

Divergences as functions of probability distributions have not been used in MI learning, due to the lack of a probability function space defined for the BS paradigm, despite the benefit of analysis independent of specific data sets [48]. Cheplygina et al. [14] proposed using the Cauchy-Schwarz divergence with a Gaussian kernel, but as a function of the instances in the bag-to-bag setting. The KL information [29] is a non-symmetric f-divergence, often used in both statistics and computer science, and is defined as follows for two probability density functions (PDFs) $f_k(\mathbf{x})$ and $f_\ell(\mathbf{x})$:

$$D_{KL}(f_k, f_\ell) = \int f_k(\mathbf{x}) \log \frac{f_k(\mathbf{x})}{f_\ell(\mathbf{x})} d\mathbf{x}. \tag{4}$$

An example of a symmetric f-divergence is the Bhattacharyya (BH) distance, defined as

$$D_{BH}(f_k, f_\ell) = -\log \int \sqrt{f_k(\mathbf{x}) f_\ell(\mathbf{x})} d\mathbf{x}, \tag{5}$$

and can be a better choice if the absolute difference, and not the ratio, differentiates the two PDFs. The appropriate divergence for a specific task can be chosen based on identified properties, e.g., for clustering [49], or a new dissimilarity function can be proposed [50].

3.5. Bag-to-Class Dissimilarity

Bag-to-bag classification can be thought of as model selection: Two bags from the training set, $f_k(\mathbf{x})$ and $f_\ell(\mathbf{x})$ are the models, and unlabeled bag $f_{bag}(\mathbf{x})$ is the sample distribution, and is labeled according to which model it resembles the most. The log-ratio test is the most powerful for model selection under certain conditions (Neyman–Pearson lemma). It is possible then to perform the log-ratio test between $f_{bag}(\mathbf{x})$ and each of the bags in the training set.

The training set in MI learning is the instances, since the bag distributions are unknown. Under the assumption that the instances from each bag are i.i.d. samples, the KL information has a special role in model selection, both from the frequentist and the Bayesian perspective. Let $f_{bag}(\mathbf{x})$ be the sample distribution (unlabeled bag), and let $f_k(\mathbf{x})$ and $f_\ell(\mathbf{x})$ be two models (labeled bags). Then the expectation over $f_{bag}(\mathbf{x})$ of the log-ratio of the two models, $E\{\log(f_k(\mathbf{x})/f_\ell(\mathbf{x}))\}$, is equal to $D_{KL}(f_{bag}, f_\ell) - D_{KL}(f_{bag}, f_k)$. In other words, the log-ratio test reveals the model closest to the sampling distribution in terms of KL information [51]. From the Bayesian viewpoint, the Akaike Information Criterion (AIC) reveals the model closest to the data in terms of KL information, and is asymptotically equivalent to Bayes factor under certain assumptions [52].

An obstacles arises: The core of MI learning is that bags from the same class are not equal, e.g., two images of the sea, so that the model is most likely not in the training set. In fact, for probability distributions with continuous parameters, the probability of the new bag being in the training set is zero. For ratio-based divergences, such as the f-divergences, the difference between $D(f_{bag}, f_k)$ and $D(f_{bag}, f_\ell)$ becomes arbitrary. Despite their necessary properties as dissimilarity measures, and the KL information's property as most powerful model selector, we see that they can fail in practice.

If the bag sampling is sparse, the dissimilarity between $f_{bag}(\mathbf{x})$ and the labeled bags becomes somewhat arbitrary regarding the true label of $f_{bag}(\mathbf{x})$. The risk is high for ratio-based divergences such as the KL information, since $f_k(\mathbf{x})/f_\ell(\mathbf{x}) = \infty$ for $\{\mathbf{x} : f_\ell(\mathbf{x}) = 0, f_k(\mathbf{x}) > 0\}$. The bag-to-bag KL information is asymptotically the best choice of divergence function, but this is not the case for sparse training sets. Bag-to-class dissimilarity makes up for some of the sparseness by aggregation of instances. Consider an image segment of color *deep green*, which appears in *sea* images, but not in *desert* images, and a segment of color *white*, which appears in both classes (waves and clouds). If the combination *deep green* and *white* does not appear in the training set, then a bag-to-bag KL information will result in infinite dissimilarity for all bags, regardless of class, but the bag-to-class KL information will be finite for the *sea* class.

Let $P(X|C) = \int_B P(X|B)dP_B(B|C)$ be the probability distribution of a random vector from the bags of class C. Let $D(P(X|B), P(X|pos))$ and $D(P(X|B), P(X|neg))$ be the divergences between the unlabeled bag and each of the classes. Choice of divergence is not obvious, since $P(X|B)$ is different from both $P(X|pos)$ and $P(X|neg)$, but can be done by identification of properties.

4. Properties for Bag-Level Classification

4.1. Properties for Bag-to-Class Divergences

We argue that the equality, orthogonality and monotonicity properties possessed by f-divergences are reasonable also for bag-to-class divergences, although less likely to occur in practice:

The equality property and the monotonicity property are valid for uncertain objects, but in practice it can occur with sparse class sampling, and we therefore argue that these properties are valid also for bag-to-class divergences. The opposite implies that a bag can be regarded more similar to one class, even though its probability distribution is identical to the other class (equality), or that, e.g., if $P_{bag}(X)$, $P_{pos}(X)$ and $P_{neg}(X)$ are Gaussian distributions with the same variance and means $\theta_{bag} < \theta_{pos} < \theta_{neg}$, we can have that $D(P(X|\theta_{bag}), P(X|\theta_{pos})) > D(P(X|\theta_{bag}), P(X|\theta_{neg}))$. In other words, we can have that the divergence between the bag and the positive class is larger than between the bag and the negative class, although the bag mean is closer to the positive class mean. This is clearly not appropriate for a dissimilarity measure.

The orthogonality property is reasonable for bag-to-class divergences: If there is no common sample space between bag and class, the divergence should take its maximum. In conclusion, f-divergences is the correct group for bag-to-class divergences.

There may be other desirable properties for bag-to-class divergences, where the aim is no longer to compare an i.i.d. sample to a model, but to compare an i.i.d. sample to an aggregation of models where the sample comes from one of them. We here propose two properties for bag-to-class divergences regarding infinite bag-to-class ratio and zero instance probability. Denote the divergence between an unlabeled bag and the reference distribution, $P_{ref}(X)$, by $D(P_{bag}, P_{ref})$.

In the *sea* images example, the class contains all possible colors that the sea can have, whereas a bag consists only of the colors of that particular moment in time. If the bag contains something that the class does not, e.g., brown color, this should be reflected in a larger divergence. However, the class should be allowed to contain something that the bag does not without this resulting in a similarly large divergence.

As a motivating example, consider the following: A positive bag, $P(X|a)$, is a continuous uniform distribution $\mathcal{U}(a, a+\delta)$, sampled according to $P(A) = \mathcal{U}(\eta, \zeta - \delta)$:

$$X|a \sim \mathcal{U}(a, a+\delta)$$
$$A \sim \mathcal{U}(\eta, \zeta - \delta)$$

A negative bag, $P(X|a')$, is $\mathcal{U}(a', a'+\delta')$ sampled according to $P(A') = \mathcal{U}(\eta', \zeta' - \delta')$:

$$X|a' \sim \mathcal{U}(a', a'+\delta')$$
$$A' \sim \mathcal{U}(\eta', \zeta' - \delta'),$$

and let $\eta' < \zeta$ so that there is an overlap between the two classes. For both positive and negative bags, we have that $P_{pos}(X)/P_{bag}(X) = \infty$ for a subspace of $\mathcal{X}$ and $P_{neg}(X)/P_{bag}(X) = \infty$ for a different subspace of $\mathcal{X}$, merely reflecting that the variability in instances within a class is larger than within a bag, as illustrated in Figure 4.

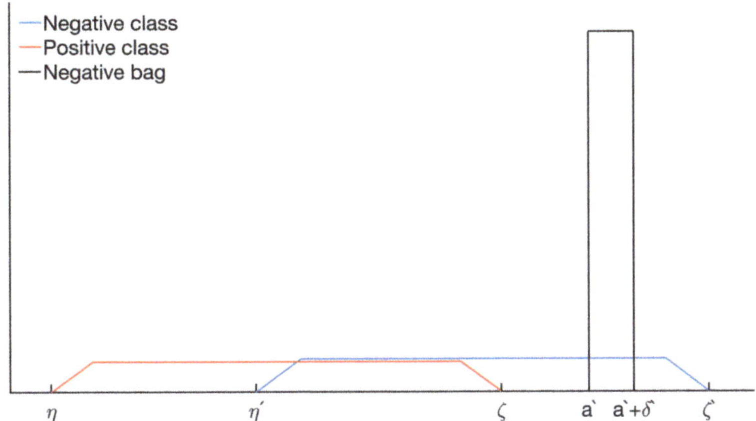

Figure 4. The PDF of a bag with uniform distribution and the PDFs of the two classes.

If $P_{bag}(X)$ is a sample from the negative class, and $P_{bag}(X)/P_{pos}(X) = \infty$ for some subspace of $\mathcal{X}$ it can easily be classified. From the above analysis, large bag-to-class ratio should be reflected in large divergence, whereas large class-to-bag ratio should not.

Property 1: Let $\mathcal{X}_M$ be the subspace of $\mathcal{X}$ where the bag-to-class ratio is larger than some M:

$$\mathcal{X}_M \subset \mathcal{X} : P_{bag}(X)/P_{ref}(X) > M,$$

and let $\mathcal{X} \setminus \mathcal{X}_M$ be its complement. Let $D^{\mathcal{X}_M}(P_{bag}, P_{ref})$ be the contribution to the total divergence for that subspace: $D(P_{bag}, P_{ref}) = D^{\mathcal{X}_M}(P_{bag}, P_{ref}) + D^{\mathcal{X} \setminus \mathcal{X}_M}(P_{bag}, P_{ref})$. Let $\mathcal{X}_M^*$ be the subspace of $\mathcal{X}$ where the class-to-bag ratio is larger than some M:

$$\mathcal{X}_M^* \subset \mathcal{X} : P_{ref}(X)/P_{bag}(X) > M,$$

and let $\mathcal{X} \setminus \mathcal{X}_M^*$ be its complement. Let $D^{\mathcal{X}_M^*}(P_{bag}, P_{ref})$ be the contribution to the total divergence for that subspace: $D(P_{bag}, P_{ref}) = D^{\mathcal{X}_M^*}(P_{bag}, P_{ref}) + D^{\mathcal{X} \setminus \mathcal{X}_M^*}(P_{bag}, P_{ref})$.

$D^{\mathcal{X}_M}$ approaches the maximum contribution as $M \to \infty$. $D_{\mathcal{X}_M^*}$ does not approach the maximum contribution as $M \to \infty$:

$$M \to \infty : \begin{cases} D^{\mathcal{X}_M}(P_{bag}, P_{ref}) \to \max(D^{\mathcal{X}_M}(P_{bag}, P_{ref})) \\ D^{\mathcal{X}_M^*}(P_{bag}, P_{ref}) \nrightarrow \max(D^{\mathcal{X}_M^*}(P_{bag}, P_{ref})). \end{cases}$$

Property 1 cannot be fulfilled by a symmetric divergence. This property is necessary in cases where the sample space of a bag is a subset of the sample space of the class, $\mathcal{X}_{bag} \subset \mathcal{X}_{class}$, e.g., for uniform distributions, and in cases where the variance of a bag is smaller than the variance of the class.

Consider $\mathcal{X}_M^*$. Because $P(X) < \infty$, this occurs for the subspace of $\mathcal{X}$ where $P_{bag}(X)$ is smaller than some ϵ and $P_{ref}(X)$ is not. We argue that when $P_{bag}(X) < \epsilon$, there should be no contribution to the divergence due to the very nature of MI learning: a bag is not a representation of the entire class, but only a small part of it.

Consider an unlabeled image coming from the class *sea*, and a binary classification problem with *desert* as the alternative class. If the unlabeled image contains only blue and white colors, it should not influence the divergence how the different shades of brown or green are distributed in the two classes, as it does not influence the likelihood of this bag coming from one class or the other. This is in contrast to bag-to-bag divergences, where this indicates a bad sample-model match.

As a second motivating example, consider the same positive class as before, and the two alternative negative classes defined by:

$$
A' \sim \begin{cases} P(A' = \eta') = 0.5 \\ P(A' = \eta' + 2\delta') = 0.5 \end{cases}
\qquad
A' \sim \begin{cases} P(A' = \eta') = 0.5 \\ P(A' = \eta' + 2\delta') = 0.25 \\ P(A' = \eta' + 3\delta') = 0.25. \end{cases}
$$

For bag classification, the question becomes: from which class is a specific bag sampled? It is equally probable that a bag $P_{\eta'}(X) = P(X|A' = \eta')$ comes from each of the two negative classes, since $P_{neg}(X)$ and $P_{neg'}(X)$ only differ where $P_{\eta'}(X) = 0$, and we argue that $D(P_{\eta'}, P_{neg})$ should be equal to $D(P_{\eta'}, P_{neg'})$.

Property 2: Let $\mathcal{X}_\epsilon$ be the subspace of $\mathcal{X}$ where $P_{bag}(X)$ is larger than some $\epsilon > 0$:

$$
\mathcal{X}_\epsilon \subset \mathcal{X} : P_{bag}(X) > \epsilon,
$$

and let $\mathcal{X} \setminus \mathcal{X}_\epsilon$ be its complement. Let $D^{\mathcal{X}_\epsilon}(P_{bag}, P_{ref})$ be the contribution to the total divergence for that subspace: $D(P_{bag}, P_{ref}) = D^{\mathcal{X}_\epsilon}(P_{bag}, P_{ref}) + D^{\mathcal{X} \setminus \mathcal{X}_\epsilon}(P_{bag}, P_{ref})$.

The contribution to the total divergence approaches zero as $\epsilon \to 0$:

$$
\epsilon \to 0 : D^{\mathcal{X}_\epsilon}(P_{bag}, P_{ref}) \to 0.
$$

This property is necessary when the bag distributions are mixture distributions with possibly zero mixture proportion. It also covers the case when the bags are different distributions, not merely have different parameters, which can be modelled as a mixture of all possible distributions in the class and only one non-zero mixture proportion.

KL information is the only divergence that fulfils these two properties among the non-symmetric divergences listed in [53]. See Appendix A. As there is no complete list of divergences, it is possible that other divergences that the authors are not aware of fulfil these properties.

4.2. A Class-Conditional Dissimilarity for MI Classification

In the *sea* and *desert* images example, consider an unlabeled image with a *pink* segment, e.g., a boat. If *pink* is absent in the training set, then the bag-to-class KL information will be infinite for both classes. We therefore propose the following property:

Property 3: For the subspace of $\mathcal{X}$ where the alternative class probability, $P_{ref'}$, is smaller than some ϵ', the contribution to the total divergence, $D_{\mathcal{X}_{\epsilon'}}$, approaches zero as $\epsilon' \to 0$:

Let $\mathcal{X}_{\epsilon'}$ be the subspace of $\mathcal{X}$ where $P_{ref'}(X)$ is larger than some $\epsilon' > 0$:

$$
\mathcal{X}_{\epsilon'} \subset \mathcal{X} : P_{ref'}(X) > \epsilon',
$$

and let $\mathcal{X} \setminus \mathcal{X}_{\epsilon'}$ be its complement. Let $D^{\mathcal{X}_{\epsilon'}}(P_{bag}, P_{ref}|P_{ref'})$ be the contribution to the total divergence for that subspace: $D(P_{bag}, P_{ref}|P_{ref'}) = D^{\mathcal{X}_{\epsilon'}}(P_{bag}, P_{ref}|P_{ref'}) + D^{\mathcal{X} \setminus \mathcal{X}_{\epsilon'}}(P_{bag}, P_{ref}|P_{ref'})$.

The contribution to the total divergence approaches zero as $\epsilon' \to 0$:

$$
\epsilon' \to 0 : D^{\mathcal{X}_{\epsilon'}}(P_{bag}, P_{ref}|P_{ref'}) \to 0.
$$

We present a class-conditional dissimilarity that accounts for this:

$$
cKL(f_{bag}, f_{pos}|f_{neg}) = \int \frac{f_{neg}(\mathbf{x})}{f_{pos}(\mathbf{x})} f_{bag}(\mathbf{x}) \log \frac{f_{bag}(\mathbf{x})}{f_{pos}(\mathbf{x})} d\mathbf{x}, \tag{6}
$$

which also fulfils Properties 1 and 2, see Appendix A.

4.3. Bag-Level Divergence Classification

With a proper dissimilarity measure, the classification task is, in theory, straightforward: A bag is given the label of its most similar class. With dense bag and class sample, the KL bag-to-bag classifier is the most powerful. There are, however, a couple of practical obstacles: The distributions from where the instances have been drawn are not known, and must be estimated. The divergences usually do not have analytical solutions, and must therefore be approximated.

We propose two similar methods based on either the ratio of bag-to-class divergences, $rD(f_{bag}, f_{pos}, f_{neg}) = D(f_{bag}, f_{pos}))/D(f_{bag}, f_{neg})$, or the class-conditional dissimilarity in Equation (6). We propose using the KL information (Equation (4)), and report for the BH distance (Equation (5)) for comparison, but any divergence function can be applied.

Given a training set $\{(\mathbb{X}_1, y_1), \ldots, (\mathbb{X}_k, y_k), \ldots, (\mathbb{X}_K, y_K)\}$ and a set, $\mathbb{X}_{bag}$, of instances drawn from an unknown distribution, $f_{bag}(\mathbf{x})$, with unknown class label y_{bag}, and let $\mathbb{X}_{neg}$ denote the set of all $\mathbf{x}_{ik} \in (\mathbb{X}_k, y_k = neg)$ and $\mathbb{X}_{pos}$ denote the set of all $\mathbf{x}_{ik} \in (\mathbb{X}_k, y_k = pos)$. The bag-level divergence classification follows the steps:

1. Estimate pdfs: Fit $\hat{f}_{neg}(\mathbf{x})$ to $\mathbb{X}_{neg}$, $\hat{f}_{pos}(\mathbf{x})$ to $\mathbb{X}_{pos}$, and $\hat{f}_{bag}(\mathbf{x})$ to $\mathbb{X}_{bag}$.

2. Calculate divergences: $D(\hat{f}_{bag}, \hat{f}_{neg}))$ and $D(\hat{f}_{bag}, \hat{f}_{pos})$,

   or $cKL(\hat{f}_{bag}, \hat{f}_{pos}|\hat{f}_{neg})$ by integral approximation.

3. Classify according to:

$$
y_{bag} = \begin{cases} pos \text{ if } rD(\hat{f}_{bag}, \hat{f}_{pos}, \hat{f}_{neg}) < t \\ neg \text{ otherwise.} \end{cases}
$$

   or

$$
y_{bag} = \begin{cases} pos \text{ if } cKL(\hat{f}_{bag}, \hat{f}_{pos}|\hat{f}_{neg}) < t \\ neg \text{ otherwise.} \end{cases}
$$

$$(7)$$

Common methods for PDF estimation are Gaussian mixture models (GMMs) and kernel density estimation (KDE). The integrals in step 2 are commonly approximated by importance sampling and Riemann sums. In rare cases, e.g., when the distributions are Gaussian, the divergences can be calculated directly. The threshold t can be pre-defined based on, e.g., misclassification penalty and prior class probabilities, or estimated from the training set by leave-one-out cross-validation. When the feature dimension is high and the number of instances in each bag is low, PDF estimation becomes arbitrary. A solution is to estimate separate PDFs for each dimension, calculate the corresponding divergences $D_1, \ldots, D_{Dim}$, and treat them as inputs into a classifier replacing step 3.

In image analysis, it has become more and more common that MI data sets are limited by the number of (labeled) bags per class, more than the number of instances per bag. With the proposed algorithm, the PDF estimates improve with increasing number of instances, and the aggregation of class instances allows for sparser bag samples.

5. Experiments

5.1. Simulated Data and Class Sparsity

The following study exemplifies the difference between BH distance ratio, rBH, KL information ratio, rKL, and cKL as classifiers for sparse training data. We investigate how the three divergences vary in accordance with the number of bags in the training set. The minimum dissimilarity bag-to-bag classifiers are also implemented, based on KL information and BH distance. The number of instances from each bag is 50, the number of bags in the training set is varied from 1 to 25 from each class, and the number of bags in the test set is 100. Each bag and its instances are sampled as described in Equation (3), and the area under the receiver operating characteristic (ROC) curve (AUC) serves as performance measure. For simplicity, we use Gaussian distributions in one dimension for *Sim 1-Sim 4*:

$$X^- \sim \mathcal{N}(\mu^-, \sigma^{2-}) \qquad\qquad X^+ \sim \mathcal{N}(\mu^+, \sigma^{2+})$$
$$\mu^- \sim \mathcal{N}(0, 10) \qquad\qquad \mu^+ \sim \mathcal{N}(\nu^+, 10)$$
$$\sigma^{2-} = |\zeta^-|, \ \zeta^- \sim \mathcal{N}(1, 1) \qquad\qquad \sigma^{2+} = |\zeta^+|, \ \zeta^+ \sim \mathcal{N}(\eta^+, 1)$$
$$\Pi^- = \pi^- \qquad\qquad \Pi^+ = 0.10.$$

Sim 1: $\nu^+ = 15$, $\eta^+ = 1$, $\pi^- = 0$: No positive instances in negative bags.
Sim 2: $\nu^+ = 15$, $\eta^+ = 1$, $\pi^- = 0.01$: Positive instances in negative bags.
Sim 3: $\nu^+ = 0$, $\eta^+ = 100$, $\pi^- = 0$: Positive and negative instances have the same expectation of the mean, but unequal variance.
Sim 4: $P(\nu^+ = 15) = P(\nu^+ = -15) = 0.5$, $\eta^+ = 1$, $\pi^- = 0.01$: Positive instances are sampled from two distributions with unequal mean expectation.

We add *Sim 5* and *Sim 6* for the discussion on instance labels in Section 6, as follows: *Sim 5* is an uncertain object classification, where the positive bags are lognormal densities with $\mu = \log(10)$ and $\sigma^2 = 0.04$, and negative bags are Gaussian mixtures densities with $\mu_1 = 9.5$, $\mu_2 = 13.5$, $\sigma^2 = 2.5$, and $\pi_1 = 0.9$. These two densities are nearly identical, see [54], p. 15. In *Sim 6*, the parameters of *Sim 5* are i.i.d. observations from Gaussian distributions, each with $\sigma^2 = 1$ for the Gaussian mixture, and $\sigma^2 = 0.04$ for the lognormal distribution. Figure 5 shows the estimated class densities and two estimated bag densities for *Sim 2* with 10 negative bags in the training set.

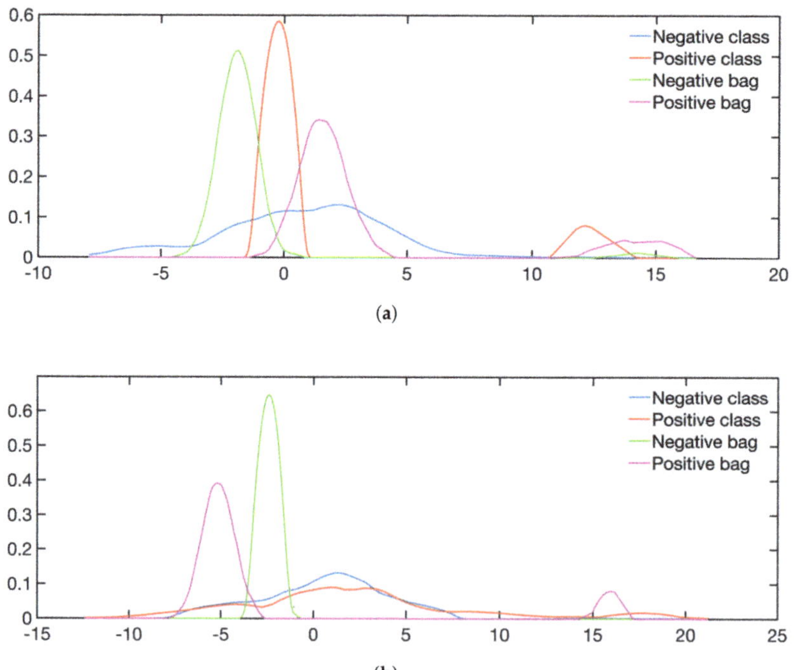

Figure 5. (**a**) One positive bag in the training set gives small variance for the class PDF. (**b**) Ten positive bags in the training set, and the variance has increased.

We use the following details for the algorithm in (7): 1. KDE fitting: Epanechnikov kernel with estimated bandwidth varying with the number of observations. 2. Integrals: Importance sampling. 3. Classifier: t is varied to give the full range of sensitivities and specificities necessary to calculate AUC.

Table 1 shows the mean AUCs for 50 repetitions.

Table 1. AUC·100 for simulated data.

| | **Bags** | | *neg:* **5** | | | *neg:* **10** | | | *neg:* **25** | |
Sim:	*pos:*	*rBH*	*rKL*	*cKL*	*rBH*	*rKL*	*cKL*	*rBH*	*rKL*	*cKL*
	1	61	69	85	62	72	89	61	73	92
1	5	63	75	86	64	82	94	68	84	97
	10	69	86	87	73	91	95	75	91	98
	1	57	61	75	59	61	78	58	55	75
2	5	59	67	79	60	68	84	62	63	85
	10	64	77	80	66	78	86	68	72	86
	1	51	55	71	52	58	73	50	57	74
3	5	53	61	76	53	66	81	52	65	83
	10	58	73	78	58	76	84	57	76	87
	1	55	61	70	56	62	73	56	58	69
4	5	56	63	75	57	64	81	59	59	80
	10	60	74	77	62	76	85	63	69	84
	1	64	61	62	67	63	66	64	62	67
5	5	73	69	63	74	70	67	75	71	72
	10	74	70	62	75	73	69	76	74	72
	1	68	68	67	66	68	68	68	71	68
6	5	65	64	67	68	68	69	70	71	74
	10	66	64	66	70	69	72	72	73	74

5.2. The Impact of Pdf Estimation and Comparison to Other Methods

We use a public data set from UCSB Center for Bio-Image Informatics to demonstrate the impact of PDF estimation method and for comparison with other MI classification methods. The UCSB data set consists of 58 breast tumor histology images, as seen in Figure 1). There are 32 images labeled as benign and 26 as malignant. The image patches are of size 7×7 pixels, and 708 features have been extracted from each patch. The mean number of instances per bag is 35. We have used the published instance values [14] to minimize other sources of variation than the classification algorithms. Following the procedure in [3], the principal components are used for dimension reduction, and 4-fold cross-validation is used so that $\hat{f}_{neg}(x)$ and $\hat{f}_{pos}(x)$ are fitted only to the instances in the training folds. Table 2 shows the AUC for rKL and cKL for three different methods for PDF estimation. GMMs are fitted to the first principal component, using an EM-algorithm, with number of components chosen by minimum AIC. In addition, KDE as in Section 5.1, and KDE with Gaussian kernel and optimal bandwidth [55] is used.

Table 2. AUC·100 for USCB breast tissue images.

	KDE (Epan.)	**KDE (Gauss.)**	**GMMs**
cKL	90	92	94
rKL	82	92	96

Table 3 shows the AUC of the GMM fitted rKL and cKL compared to four other MI learning methods. For articles presenting more than one method, the best-performing method is displayed in Table 3.

Table 3. AUC·100 for USCB breast tissue images.

Method	AUC
cKL	94
rKL	96
DEEPISR-MIL [34]	90
Li et al. [33]	93
GPMIL [3]	86
RGPMIL [3]	90

5.3. Comparison to State-of-the-Art Methods

The benchmark data sets that have been used for comparison of MIL methods have particularly low number of instances compared to the number of features. e.g., in *Musk*1, more than half of the bags contain less than 5 instances, and in *Musk*2, one fourth of the bags contain less than 5 instances. It is obvious that a PDF-based method will not work. The COREL data base, previously used in MIL method comparisons, is no longer available, only data sets with extracted features. Again, the number of instances is too low for density estimation. In addition, [56] showed how the feature extraction methods influence the results of MIL classifiers.

We here present the results of *cKL* and *rKL* compared to the five best-performing MIL methods using the *BreakHis* data set, as presented in [31]. This data set is suited for PDF-based methods, since the images themselves are available, and hence, the number of instances can be adjusted to assure a sufficiently dense sampling. We follow the procedure in [31], using the 162 parameter-free threshold adjacency statistics (PFTAS)

features for 1000 image patches of size 64×64. Dimension reduction is done by principal components, so that 90% of the variance is explained, and the dimension is reduced to about 25, depending on which data set, see Table 4. Each data set is split into training, validation and test sets (35%/35%/30%), where we use the exact same five test sets as [31]. There are multiple images from the same tumor, but the data set is split so that the same tumor does not appear in both training/validation and test set.

We use the following details for the algorithm in (7):

1. GMMs are fitted with $1, \ldots, 100$ components, and the number of components is chosen by minimum AIC. To save computation time, the number of components is estimated for 10 bags sampled from the training set. The median number of components is used to fit the bag PDFs in the rest of the algorithm, see Table 4. For the class PDFs, a random subsample of 10% of the instances is taken from each bag, to reduce computation time.
2. Integrals: Importance sampling.
3. Classification: To estimate the threshold, t, the training set is used to estimate $f_{pos}^{train}(\mathbf{x})$ and $f_{neg}^{train}(\mathbf{x})$, and the divergences between the bags in the validation set and $f_{pos}^{train}(\mathbf{x})$ and $f_{neg}^{train}(\mathbf{x})$ are calculated. The threshold, $\hat{t}$, that gives the highest accuracy will then serve as threshold for the test set.

Please note that the bags from the test set is not involved in picking the number of components or estimating $\hat{t}$.

Table 4. Number of components.

Data Set	40$\times$	100$\times$	200$\times$	400$\times$
Dimension	23	26	25	24
Rep 1	66	55	52	70
Rep 2	58	49	69	71
Rep 3	59	50	50	70
Rep 4	47	49	58	73
Rep 5	63	59	72	74

5.4. Results

The general trend in Table 1 is that *cKL* gives higher AUC than *rKL*, which in turn gives higher AUC than *rBH*, in line with the divergences' properties for sparse training sets. The same trend can be seen with a Gaussian kernel and optimal bandwidth (numbers not reported). The gap between *cKL* and *rKL* narrows with larger training sets. In other words, the benefit of *cKL* increases with sparsity. This can be explained by the ∞/∞ risk of *rKL*, as seen in Figure 5a. Increasing π^+ also narrows the gap between *rKL* and *cKL*, and eventually (at approximately $\pi^+ = 0.25$), *rKL* outperforms *cKL* (numbers not reported). *Sim 1* and *Sim 3* are less affected because the ratio π^+/π^- is already ∞.

The minimum bag-to-bag classifier gives a single sensitivity-specificity outcome, and the KL information outperforms the BH distance. Compared to the ROC curve, as illustrated in Figure 6, the minimum bag-to-bag KL information classifier exceeds the bag-to-class dissimilarities only for very large training sets, typically for 500 or more, then at the expense of extensive computation time.

Sim 5 is an example in which the absolute difference, not the ratio, differentiates the two classes, and *rBH* has the superior performance. When the extra hierarchy level is added in *Sim 6*, the performances returned to normal.

The UCSB breast tissue study shows that the simple divergence-based approach can outperform more sophisticated algorithms. *rKL* is more sensitive than *cKL* to choice of density estimation method, as shown in Table 2. *rKL* performs better than *cKL* with GMM, and both are among the best performing in Table 3. The study is too small to draw conclusions. Table 2 shows how the performance can vary between two common PDF estimation methods that do not assume a particular underlying distribution. Both KDE and GMM are sensitive to chosen parameters or parameter estimation method, bandwidth and number of components, respectively, and no method will fit all data sets. In general, KDE is faster, but more sensitive to bandwidth, whereas GMM is more stable. For bags with very few instances the benefits of GMM cannot be exploited, and KDE is preferred.

The BreakHis study shows that both *rKL* and *cKL* perform as good as or better than the other methods, the exception being *cKL* for 40$\times$, as reported in Table 5. "As good as" refers to the mean being within one standard deviation of the highest mean. Since none of the methods have overall superior performance, we believe that the differences within one standard deviation is not enough to declare a winner. *rKL* has overall best performance in the sense that it is always within one standard deviation from the highest mean. However, *cKL*, *MI-SVM poly* and *Non-parametric* follow close behind with four out of five. Therefore, we will again avoid declaring a winner. Table 4 demonstrates that the number of components varies between repetitions, but does not influence the accuracy substantially. For reference, we have reported the AUC in Table 6, as this is a common way of reporting performance in the MIL context.

Table 5. Accuracy and standard deviation. Best results and those within one standard deviation in bold.

Data Set (Magnification)	40×	100×	200×	400×
MI-SVM poly [57]	**86.2** (2.8)	**82.8** (4.8)	81.7 (4.4)	**82.7** (3.8)
Non-parametric [58]	**87.8** (5.6)	**85.6** (4.3)	80.8 (2.8)	**82.9** (4.1)
MILCNN [59]	**86.1** (4.2)	**83.8** (3.1)	80.2 (2.6)	80.6 (4.6)
CNN [31]	**85.6** (4.8)	**83.5** (3.9)	83.1 (1.9)	80.8 (3.0)
SVM [31]	79.9 (3.7)	77.1 (5.5)	84.2 (1.6)	81.2 (3.6)
rKL	**83.4** (4.1)	**84.9** (4.2)	**88.3** (3.6)	**84.0** (2.8)
cKL	81.5 (3.2)	**85.2** (3.5)	**88.1** (3.6)	**85.0** (3.5)

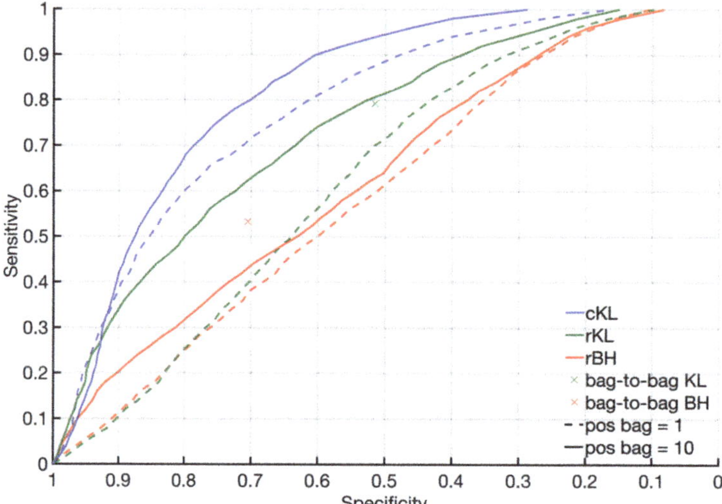

Figure 6. An example of ROC curves for *cKL*, *rKL* and *rBH* classifiers. The performance increases when the number of positive bags in the training set increases from 1 (dashed line) to 10 (solid line). The sensitivity-specificity pairs for the bag-to-bag KL and BH classifier is displayed for 100 positive and negative bags in the training set for comparison.

Table 6. AUC and standard deviation.

Data Set (Magnification)	40×	100×	200×	400×
rKL	91.4 (2.4)	91.3 (2.2)	94.4 (1.9)	91.6 (1.7)
cKL	88.4 (2.6)	89.7 (1.6)	91.9 (2.7)	91.7 (2.4)

The superior performance of *cKL* for the KDE (Epan.) in Table 2 can be explained by the Epanechnikov kernel's zero value, as opposed to the Gaussian kernel which is always positive. *rKL* will then suffer from its ∞/∞ property given the limited training set for each class. With Gaussian kernel and GMMs, *rKL* improves its performance compared to *cKL*, as demonstrated in the simulation study. For the BreakHist data, *rKL* and *cKL* show similar performance. Although *cKL* is not within one standard deviation from the best-performing method for the 40× data set, it is within one standard deviation from *rKL*. The similar performance of *rKL* and *cKL* is in line with the simulation study where the superiority of *cKL* is demonstrated for sparse training sets, but not for all types of data.

6. Discussion

6.1. Point-of-View

The theoretical basis of the bag-to-class divergence approach relies on viewing a bag as a probability distribution, hence fitting into the branch of collective assumptions of the Foulds and Frank taxonomy [13]. The probability distribution estimation can be seen as extracting bag-level information from a set $\mathbb{X}$, and hence falls into the BS paradigm of Amores [15]. The probability distribution space is non-vectorial, different from the distance-kernel spaces in [15], and divergences are used for classification.

In practice, the evaluation points of the importance sampling gives a mapping from the set $\mathbb{X}$ to a single vector, $\hat{f}_{bag}(\mathbf{z})$. The mapping concurs with the ES paradigm, and the same applies for the graph-based methods. From that viewpoint, the bag-to-class divergence approach expands the distance branch of Foulds and Frank to include a bag-to-class category in addition to instance-level and bag-level distances. However, the importance sampling is a technicality of the algorithm. We argue that the method belongs to the BS paradigm. When the divergences are used as input to a classifier, the ES paradigm is a better description.

Carbonneau et al. [16] assume underlying instance labels. From a probability distribution viewpoint, this corresponds to posterior probabilities, which are in practice, inaccessible. In *Sim 1–Sim 4*, the instance labels are inaccessible through observations without previous knowledge about the distributions. In *Sim 6*, the instance label approach is not useful due to the similarity between the two distributions:

$$
\begin{aligned}
X|\theta^+ &\sim P(X|\theta^+) & X|\theta^- &\sim P(X|\theta^-) \\
\Theta^+ &\sim P(\Theta^+) & \Theta^- &\sim P(\Theta^-),
\end{aligned}
\tag{8}
$$

where $P(X|\Theta^+)$ and $P(X|\Theta^-)$ are the lognormal and the Gaussian mixture, respectively. Equation (3) is just a special case of Equation (8), where Θ^+ is the random vector $\{\Theta, \Pi_{pos}\}$. Without knowledge about the distributions, discriminating between training sets following the generative model of Equations (3) and (8) is only possible for a limited number of problems. Even the uncertain objects of *Sim 5* are difficult to discriminate from MI objects based solely on the observations in the training set.

6.2. Conclusions and Future Work

Although the bag-to-bag KL information has the minimum misclassification rate, the typical bag sparseness of MI training sets is an obstacle. This is partly solved by bag-to-class dissimilarities and the proposed class-conditional KL information accounts for additional sparsity of bags.

The bag-to-class divergence approach addresses three main challenges of MI learning. (1) Aggregation of instances according to bag label and the additional class-conditioning provide a solution for the bag sparsity problem. (2) The bag-to-bag approach suffers from extensive computation time, solved by the bag-to-class approach. (3) Viewing bags as probability distributions give access to analytical tools from statistics and probability theory, and comparisons of methods can be done on a data-independent level through identification of properties. The properties presented here are not an extensive list, and any extra knowledge should be taken into account whenever available.

A more thorough analysis of the proposed function, *cKL*, will identify its weaknesses and strengths, and can lead to improved versions as well as alternative class-conditional dissimilarity measures and a more comprehensive tool.

The diversity of data types, assumptions, problem characteristics, sampling sparsity, etc. is far too large for any one approach to be sufficient. The introduction of divergences as an alternative class of dissimilarity functions, and the bag-to-class dissimilarity as an alternative to the bag-to-bag dissimilarity, has added additional tools to the MI toolbox.

Author Contributions: Conceptualization, K.M.; methodology, K.M.; software, K.M.; writing—original draft preparation, K.M.; writing—review and editing, J.Y.H. and F.G.; visualization, K.M.; supervision, J.Y.H. and F.G. All authors have read and agreed to the published version of the manuscript.

Funding: The publication charges for this article have been funded by a grant from the publication fund of UiT The Arctic University of Norway.

Conflicts of Interest: The authors declare no conflict of interest.

Abbreviations

The following abbreviations are used in this manuscript:

MI	multi-instance
PDF	probability density function
IS	instance space
ES	embedded space
BS	bag space
KL	Kullback–Leibler
SVM	support vector machine
AIC	Akaike Information Criterion
GMM	Gaussian mixture models
KDE	kernel density estimation
ROC	receiver operating characteristic
AUC	area under the ROC curve

Appendix A

For the sake of readability, we repeat summary versions of the properties here:

Property 1:

$$\mathcal{X}_M \subset \mathcal{X} : P_{bag}(X)/P_{ref}(X) > M$$

$$M \to \infty : \begin{cases} D^{\mathcal{X}_M}(P_{bag}, P_{ref}) \to \max(D^{\mathcal{X}_M}(P_{bag}, P_{ref})) \\ D^{\mathcal{X}_M^*}(P_{bag}, P_{ref}) \not\to \max(D^{\mathcal{X}_M^*}(P_{bag}, P_{ref})) \end{cases}$$

Property 2:

$$\mathcal{X}_\epsilon \subset \mathcal{X} : P_{bag}(X) > \epsilon$$

$$\epsilon \to 0 : D^{\mathcal{X}_\epsilon}(P_{bag}, P_{ref}) \to 0$$

Property 3:

$$\mathcal{X}_{\epsilon'} \subset \mathcal{X} : P_{ref'}(X) > \epsilon'$$

$$\epsilon' \to 0 : D^{\mathcal{X}_{\epsilon'}}(P_{bag}, P_{ref}|P_{ref'}) \to 0$$

Appendix A.1. Non-Symmetric Divergences:

We show that the only non-symmetric divergences listed in [53] that fulfil both Property 1 and Property 2 is the KL information. For all other divergences, we show one property that it does not fulfil.

The χ^2-divergence, defined as:

$$\int \frac{(f_{bag}(\mathbf{x}) - f_{ref}(\mathbf{x}))^2}{f_{ref}(\mathbf{x})} d\mathbf{x},$$

does not fulfil Property 2:

$$\int_{\mathcal{X}_\epsilon} \frac{(\epsilon - f_{ref}(\mathbf{x}))^2}{f_{ref}(\mathbf{x})} d\mathbf{x} \to \int_{\mathcal{X}_\epsilon} f_{ref}(\mathbf{x}) d\mathbf{x} \nrightarrow 0.$$

The KL information, referred to as Relative information in [53], defined as:

$$\int f_{bag}(\mathbf{x}) \log \frac{f_{bag}(\mathbf{x})}{f_{ref}(\mathbf{x})} d\mathbf{x},$$

fulfils Property 1:

$$\int_{\mathcal{X}_M} f_{bag}(\mathbf{x}) \log M \, d\mathbf{x} \to \infty = \max$$

$$\int_{\mathcal{X}_{M^*}} f_{bag}(\mathbf{x}) \log \frac{1}{M} \, d\mathbf{x} \nrightarrow \infty,$$

since $f_{bag}(\mathbf{x}) < \infty$ and $\frac{1}{M} < \infty$, and Property 2:

$$\int_{\mathcal{X}_\epsilon} \epsilon \log \frac{\epsilon}{f_{ref}(\mathbf{x})} d\mathbf{x} \to 0 = \min$$

The Relative Jensen-Shannon divergence, defined as:

$$\int f_{bag}(\mathbf{x}) \log \frac{2 f_{bag}(\mathbf{x})}{f_{bag}(\mathbf{x}) + f_{ref}(\mathbf{x})} d\mathbf{x},$$

does not fulfil Property 1:

$$\int_{\mathcal{X}_M} f_{bag}(\mathbf{x}) \log \frac{2}{1 + \frac{f_{ref}(\mathbf{x})}{f_{bag}(\mathbf{x})}} d\mathbf{x} = \int_{\mathcal{X}_M} f_{bag}(\mathbf{x}) \log \frac{2}{1 + \frac{1}{M}} d\mathbf{x} \to \int_{\mathcal{X}_M} f_{bag}(\mathbf{x}) \log 2 \, d\mathbf{x} = \nrightarrow \max.$$

The Relative Arithmetic-Geometric divergence, defined as:

$$\int \frac{f_{bag}(\mathbf{x}) + f_{ref}(\mathbf{x})}{2} \log \frac{f_{bag}(\mathbf{x}) + f_{ref}(\mathbf{x})}{2 f_{bag}(\mathbf{x})} d\mathbf{x},$$

does not fulfil Property 2:

$$\int_{\mathcal{X}_\epsilon} \frac{\epsilon + f_{ref}(\mathbf{x})}{2} \log \frac{\epsilon + f_{ref}(\mathbf{x})}{2\epsilon} d\mathbf{x} \to \infty \neq \min.$$

The Relative J-divergence, defined as:

$$\int (f_{bag}(\mathbf{x}) + f_{ref}(\mathbf{x})) \log \frac{f_{bag}(\mathbf{x}) + f_{ref}(\mathbf{x})}{2 f_{bag}(\mathbf{x})} d\mathbf{x},$$

does not fulfil Property 2:

$$\int_{\mathcal{X}_\epsilon} (\epsilon + f_{ref}(\mathbf{x})) \log \frac{\epsilon + f_{ref}(\mathbf{x})}{2\epsilon} d\mathbf{x} \to \infty \neq \min.$$

Appendix A.2. Class-Conditional Bag-to-Class Divergence

Class-conditional KL-divergence:

For the class-conditional divergence, there are three PDFs involved, and therefore, we have some additional restrictions. We show that the Equality and Orthogonality properties of f-divergences are fulfilled also by cKL. We were not able to conclude regarding the Monotonicity property.

$$cKL(f_{bag}, f_{pos} | f_{neg}) = \int \frac{f_{neg}(\mathbf{x})}{f_{pos}(\mathbf{x})} f_{bag}(\mathbf{x}) \log \frac{f_{bag}(\mathbf{x})}{f_{pos}(\mathbf{x})} d\mathbf{x}$$

Equality, $f_{bag}(\mathbf{x}) = f_{pos}(\mathbf{x})$, $f_{neg} \geq 0$:

$$\int \frac{f_{neg}(\mathbf{x})}{f_{pos}(\mathbf{x})} f_{pos}(\mathbf{x}) \log \frac{f_{pos}(\mathbf{x})}{f_{pos}(\mathbf{x})} d\mathbf{x} = \int f_{neg}(\mathbf{x}) \log 1 \, d\mathbf{x} = 0 = \min.$$

Orthogonality, $f_{bag}(\mathbf{x})/f_{pos}(\mathbf{x}) = \infty$, $f_{neg}(\mathbf{x}) > 0$:

$$\int f_{neg}(\mathbf{x}) \frac{f_{bag}(\mathbf{x})}{f_{pos}(\mathbf{x})} \log \frac{f_{bag}(\mathbf{x})}{f_{pos}(\mathbf{x})} d\mathbf{x} = \infty = \max.$$

Property 1: $f_{neg}(\mathbf{x}) \cdot M > 0$

$$\int_{\mathcal{X}_M} f_{neg}(\mathbf{x}) M \log M \, d\mathbf{x} \to \infty = \max$$

$$\int_{\mathcal{X}_{M^*}} \frac{f_{neg}(\mathbf{x})}{M} \log \frac{1}{M} \, d\mathbf{x} \not\to \infty$$

Property 2: $\frac{f_{neg}(\mathbf{x})}{f_{pos}(\mathbf{x})} \cdot \epsilon > 0$

$$\int_{\mathcal{X}_\epsilon} \frac{f_{neg}(\mathbf{x})}{f_{pos}(\mathbf{x})} \epsilon \log \frac{\epsilon}{f_{pos}(\mathbf{x})} d\mathbf{x} \to 0$$

Property 3: $f_{bag}(\mathbf{x})/f_{pos}(\mathbf{x}) > M$, $\epsilon' \to 0$ faster than $M \to \infty$

$$\int_{\mathcal{X}_{\epsilon'}} \frac{\epsilon'}{f_{pos}(\mathbf{x})} f_{bag}(\mathbf{x}) \log \frac{f_{bag}(\mathbf{x})}{f_{pos}(\mathbf{x})} d\mathbf{x} \to 0$$

References

1. Cheplygina, V.; de Bruijne, M.; Pluim, J.P. Not-so-supervised: A survey of semi-supervised, multi-instance, and transfer learning in medical image analysis. *Med. Image Anal.* **2019**, *54*, 280–296. doi:10.1016/j.media.2019.03.009.
2. Gelasca, E.D.; Byun, J.; Obara, B.; Manjunath, B.S. Evaluation and Benchmark for Biological Image Segmentation. In Proceedings of the IEEE International Conference on Image Processing, San Diego, CA, USA, 12–15 October 2008; pp. 1816–1819. doi:10.1109/ICIP.2008.4712130.
3. Kandemir, M.; Zhang, C.; Hamprecht, F.A. Empowering Multiple Instance Histopathology Cancer Diagnosis by Cell Graphs. In Proceedings of the International Conference on Medical Image Computing and Computer-Assisted Intervention—MICCAI 2014, Boston, MA, USA, 14–18 September 2014; pp. 228–235. doi:10.1007/978-3-319-10470-6_29.
4. Doran, G.; Ray, S. Multiple-Instance Learning from Distributions. *J. Mach. Learn. Res.* **2016**, *17*, 1–50.
5. Zhang, M.L.; Zhou, Z.H. Multi-instance clustering with applications to multi-instance prediction. *Appl. Intell.* **2009**, *31*, 47–68. doi:10.1007/s10489-007-0111-x.
6. Zhou, Z.H.; Zhang, M.L.; Huang, S.J.; Li, Y.F. Multi-instance multi-label learning. *Artif. Intell.* **2012**, *176*, 2291–2320. doi:10.1016/j.artint.2011.10.002.
7. Tang, P.; Wang, X.; Huang, Z.; Bai, X.; Liu, W. Deep patch learning for weakly supervised object classification and discovery. *Pattern Recognit.* **2017**, *71*, 446–459. doi:10.1016/J.PATCOG.2017.05.001.

8. Wang, X.; Yan, Y.; Tang, P.; Bai, X.; Liu, W. Revisiting multiple instance neural networks. *Pattern Recognit.* **2018**, *74*, 15–24. doi:10.1016/J.PATCOG.2017.08.026.

9. Dietterich, T.G.; Lathrop, R.H.; Lozano-Pérez, T. Solving the multiple instance problem with axis-parallel rectangles. *Artif. Intell.* **1997**, *89*, 31–71. doi:10.1016/s0004-3702(96)00034-3.

10. Xu, Y.Y. Multiple-instance learning based decision neural networks for image retrieval and classification. *Neurocomputing* **2016**, *171*, 826–836. doi:10.1016/j.neucom.2015.07.024.

11. Qiao, M.; Liu, L.; Yu, J.; Xu, C.; Tao, D. Diversified dictionaries for multi-instance learning. *Pattern Recognit.* **2017**, *64*, 407–416. doi:10.1016/j.patcog.2016.08.026.

12. Weidmann, N.; Frank, E.; Pfahringer, B. A Two-Level Learning Method for Generalized Multi-instance Problems. In Proceedings of the European Conference on Machine Learning, Cavtat-Dubrovnik, Croatia, 22–26 September 2003; pp. 468–479. doi:10.1007/978-3-540-39857-8_42.

13. Foulds, J.; Frank, E. A review of multi-instance learning assumptions. *Knowl. Eng. Rev.* **2010**, *25*, 1–25. doi:10.1017/s026988890999035x.

14. Cheplygina, V.; Tax, D.M.J.; Loog, M. Multiple Instance Learning with Bag Dissimilarities. *Pattern Recognit.* **2015**, *48*, 264–275. doi:10.1016/j.patcog.2014.07.022.

15. Amores, J. Multiple Instance Classification: Review, Taxonomy and Comparative Study. *Artif. Intell.* **2013**, *201*, 81–105. doi:10.1016/j.artint.2013.06.003.

16. Carbonneau, M.A.; Cheplygina, V.; Granger, E.; Gagnon, G. Multiple Instance Learning: A survey of Problem Characteristics and Applications. *Pattern Recognit.* **2018**, *77*, 329–353. doi:10.1016/j.patcog.2017.10.009.

17. Maron, O.; Lozano-Pérez, T. A framework for multiple-instance learning. In *Advances in Neural Information Processing Systems, Denver, CO, USA, 30 November–5 December 1998*; MIT Press: Cambridge, MA, USA, 1998; Volume 10, pp. 570–576.

18. Xu, X.; Frank, E. *Logistic Regression and Boosting for Labeled Bags of Instances*; Dai, H., Srikant, R., Zhang, C., Eds.; Lecture Notes in Computer Science; Springer: Berlin/Heidelberg, Germany, 2004; pp. 272–281. doi:10.1007/978-3-540-24775-3_35.

19. Tax, D.M.J.; Loog, M.; Duin, R.P.W.; Cheplygina, V.; Lee, W.J. Bag Dissimilarities for Multiple Instance Learning. In *Similarity-Based Pattern Recognition*; Lecture Notes in Computer Science; Pelillo, M., Hancock, E.R., Eds.; Springer: Berlin/Heidelberg, Germany, 2011; Volume 7005, pp. 222–234. doi:10.1007/978-3-642-24471-1_16.

20. Zhou, Z.H.; Sun, Y.Y.; Li, Y.F. Multi-instance Learning by Treating Instances As non-I.I.D. Samples. In Proceedings of the 26th Annual International Conference on Machine Learning—ICML '09, Montreal, QC, Canada, 14–18 June 2009; pp. 1249–1256. doi:10.1145/1553374.1553534.

21. Cheplygina, V.; Tax, D.M.J.; Loog, M. Dissimilarity-Based Ensembles for Multiple Instance Learning. *IEEE Trans. Neural Netw. Learn. Syst.* **2016**, *27*, 1379–1391. doi:10.1109/TNNLS.2015.2424254.

22. Boiman, O.; Shechtman, E.; Irani, M. In defense of Nearest-Neighbor based image classification. In Proceedings of the 2008 IEEE Conference on Computer Vision and Pattern Recognition, Anchorage, AK, USA, 23–28 June 2008; pp. 1–8. doi:10.1109/cvpr.2008.4587598.

23. Lee, W.J.; Cheplygina, V.; Tax, D.M.J.; Loog, M.; Duin, R.P.W. Bridging structure and feature representations in graph matching. *Int. J. Patten Recognit. Artif. Intell.* **2012**, *26*, 1260005. doi:10.1142/s0218001412600051.

24. Scott, S.; Zhang, J.; Brown, J. On generalized multiple-instance learning. *Int. J. Comput. Intell. Appl.* **2005**, *5*, 21–35. doi:10.1142/s1469026805001453.

25. Ruiz-Muñoz, J.F.; Castellanos-Dominguez, G.; Orozco-Alzate, M. Enhancing the dissimilarity-based classification of birdsong recordings. *Ecol. Inform.* **2016**, *33*, 75–84. doi:10.1016/j.ecoinf.2016.04.001.

26. Sørensen, L.; Loog, M.; Tax, D.M.J.; Lee, W.J.; de Bruijne, M.; Duin, R.P.W. Dissimilarity-Based Multiple Instance Learning. In *Structural, Syntactic, and Statistical Pattern Recognition*; Hancock, E.R., Wilson, R.C., Windeatt, T., Ulusoy, I., Escolano, F., Eds.; Springer: Berlin/Heidelberg, Germany, 2010; pp. 129–138. doi:10.1007/978-3-642-14980-1_12.

27. Schölkopf, B. The Kernel Trick for Distances. In Proceedings of the 13th International Conference on Neural Information Processing Systems, Denver, CO, USA, 27 November–2 December 2000; MIT Press: Cambridge, MA, USA, 2000; pp. 283–289.

28. Wei, X.S.; Wu, J.; Zhou, Z.H. Scalable Algorithms for Multi-Instance Learning. *IEEE Trans. Neural Netw. Learn. Syst.* **2017**, *28*, 975–987. doi:10.1109/TNNLS.2016.2519102.

29. Kullback, S.; Leibler, R.A. On Information and Sufficiency. *Ann. Math. Stat.* **1951**, *22*, 79–86. doi:10.1214/aoms/1177729694.

30. Sahu, S.K.; Cheng, R.C.H. A fast distance-based approach for determining the number of components in mixtures. *Can. J. Stat.* **2003**, *31*, 3–22. doi:10.2307/3315900.

31. Sudharshan, P.; Petitjean, C.; Spanhol, F.; Oliveira, L.E.; Heutte, L.; Honeine, P. Multiple instance learning for histopathological breast cancer image classification. *Expert Syst. Appl.* **2019**, *117*, 103–111. doi:10.1016/J.ESWA.2018.09.049.

32. Zhang, G.; Yin, J.; Li, Z.; Su, X.; Li, G.; Zhang, H. Automated skin biopsy histopathological image annotation using multi-instance representation and learning. *BMC Med. Genom.* **2013**, *6*, S10. doi:10.1186/1755-8794-6-S3-S10.

33. Li, W.; Zhang, J.; McKenna, S.J. Multiple instance cancer detection by boosting regularised trees. In *Lecture Notes in Computer Science (Including Subseries Lecture Notes in Artificial Intelligence and Lecture Notes in Bioinformatics)*; Springer: Basel, Switzerland, 2015; Volume 9349, pp. 645–652. doi:10.1007/978-3-319-24553-9_79.

34. Tomczak, J.M.; Ilse, M.; Welling, M. Deep Learning with Permutation-invariant Operator for Multi-instance Histopathology Classification. *arXiv* **2017**, arXiv:1712.00310.

35. Mercan, C.; Aksoy, S.; Mercan, E.; Shapiro, L.G.; Weaver, D.L.; Elmore, J.G. Multi-Instance Multi-Label Learning for Multi-Class Classification of Whole Slide Breast Histopathology Images. *IEEE Trans. Med. Imaging* **2018**, *37*, 316–325. doi:10.1109/TMI.2017.2758580.

36. Xu, Y.; Zhu, J.Y.; Chang, E.I.; Lai, M.; Tu, Z. Weakly supervised histopathology cancer image segmentation and classification. *Med. Image Anal.* **2014**, *18*, 591–604. doi:10.1016/j.media.2014.01.010.

37. McCann, M.T.; Bhagavatula, R.; Fickus, M.C.; Ozolek, J.A.; Kovačević, J. Automated colitis detection from endoscopic biopsies as a tissue screening tool in diagnostic pathology. In Proceedings of the 2012 19th IEEE International Conference on Image Processing, Orlando, FL, USA, 30 September–3 October 2012; pp. 2809–2812. doi:10.1109/ICIP.2012.6467483.

38. Dundar, M.M.; Badve, S.; Raykar, V.C.; Jain, R.K.; Sertel, O.; Gurcan, M.N. A multiple instance learning approach toward optimal classification of pathology slides. In Proceedings of the International Conference on Pattern Recognition, Istanbul, Turkey, 23–26 August 2010; pp. 2732–2735. doi:10.1109/ICPR.2010.669.

39. Samsudin, N.A.; Bradley, A.P. Nearest neighbour group-based classification. *Pattern Recognit.* **2010**, *43*, 3458–3467. doi:10.1016/j.patcog.2010.05.010.

40. O.Z. Kraus, J.L. Ba, B.F. Classifying and segmenting microscopy imageswith deep multiple instance learning. *Bioinformatics* **2016**, *32*, i52–i59. doi:10.1093/bioinformatics/btw252.

41. Hou, L., Samaras, D., Kurc, T. M., Gao, Y., Davis, J. E., Saltz, J.H. Efficient Multiple Instance Convolutional Neural Networks for GigapixelResolution Image Classification. *arXiv* **2015**, arXiv:1504.07947.

42. Jia, Z.; Huang, X.; Chang, E.I.; Xu, Y. Constrained Deep Weak Supervision for Histopathology Image Segmentation. *IEEE Trans. Med. Imaging* **2017**, *36*, 2376–2388. doi:10.1109/TMI.2017.2724070.

43. Jiang, B.; Pei, J.; Tao, Y.; Lin, X. Clustering Uncertain Data Based on Probability Distribution Similarity. *IEEE Trans. Knowl. Data Eng.* **2013**, *25*, 751–763. doi:10.1109/tkde.2011.221.

44. Kriegel, H.P.; Pfeifle, M. Density-based Clustering of Uncertain Data. In Proceedings of the Eleventh ACM SIGKDD International Conference on Knowledge Discovery in Data Mining KDD '05, Chicago, IL, USA, 21–24 August 2005; pp. 672–677. doi:10.1145/1081870.1081955.

45. Ali, S.M.; Silvey, S.D. A General Class of Coefficients of Divergence of One Distribution from Another. *J. R. Stat. Soc. Ser. B (Methodol.)* **1966**, *28*, 131–142.

46. Csiszár, I. Information-type measures of difference of probability distributions and indirect observations. *Studia Scientiarum Mathematicarum Hungarica* **1967**, *2*, 299–318.

47. Berger, A. On orthogonal probability measures. *Proc. Am. Math. Soc.* **1953**, *4*, 800–806. doi:10.1090/s0002-9939-1953-0056868-5.

48. Gibbs, A.L.; Su, F.E. On Choosing and Bounding Probability Metrics. *Int. Stat. Rev.* **2002**, *70*, 419–435. doi:10.1111/j.1751-5823.2002.tb00178.x.

49. Møllersen, K.; Dhar, S.S.; Godtliebsen, F. On Data-Independent Properties for Density-Based Dissimilarity Measures in Hybrid Clustering. *Appl. Math.* **2016**, *07*, 1674–1706. doi:10.4236/am.2016.715143.

50. Møllersen, K.; Hardeberg, J.Y.; Godtliebsen, F. Divergence-based colour features for melanoma detection. In Proceedings of the 2015 Colour and Visual Computing Symposium (CVCS), Gjøvik, Norway, 25–26 August 2015; pp. 1–6. doi:10.1109/CVCS.2015.7274885.

51. Eguchi, S.; Copas, J. Interpreting Kullback-Leibler Divergence with the Neyman-Pearson Lemma. *J. Multivar. Anal.* **2006**, *97*, 2034–2040. doi:10.1016/j.jmva.2006.03.007.

52. Kass, R.E.; Raftery, A.E. Bayes Factors. *J. Am. Stat. Assoc.* **1995**, *90*, 773–795. doi:10.2307/2291091.

53. Taneja, I.J.; Kumar, P. Generalized non-symmetric divergence measures and inequaities. *J. Interdiscip. Math.* **2006**, *9*, 581–599. doi:10.1080/09720502.2006.10700466.

54. McLachlan, G.; Peel, D. *Finite Mixture Models*; Wiley Series in Probability and Statistics; John Wiley & Sons, Inc.: Hoboken, NJ, USA, 2000. doi:10.1002/0471721182.

55. Sheather, S.J.; Jones, M.C. A Reliable Data-Based Bandwidth Selection Method for Kernel Density Estimation. *J. R. Stat. Soc. Ser. B (Methodol.)* **1991**, *53*, 683–690.

56. Wei, X.S.; Zhou, Z.H. An empirical study on image bag generators for multi-instance learning. *Mach. Learn.* **2016**, *105*, 155–198. doi:10.1007/s10994-016-5560-1.

57. Andrews, S.; Andrews, S.; Tsochantaridis, I.; Hofmann, T. Support vector machines for multiple-instance learning. *Adv. Neural Inf. Process. Syst.* **2003**, *15*, 561—568.

58. Venkatesan, R.; Chandakkar, P.; Li, B. Simpler Non-Parametric Methods Provide as Good or Better Results to Multiple-Instance Learning. In Proceedings of the IEEE International Conference on Computer Vision, Santiago, Chile, 7–13 December 2015.

59. Sun, M.; Han, T.X.; Ming-Chang Liu.; Khodayari-Rostamabad, A. Multiple Instance Learning Convolutional Neural Networks for object recognition. In Proceedings of the 2016 23rd International Conference on Pattern Recognition (ICPR), Cancun, Mexico, 4–8 December 2016; pp. 3270–3275. doi:10.1109/ICPR.2016.7900139.

© 2020 by the authors. Licensee MDPI, Basel, Switzerland. This article is an open access article distributed under the terms and conditions of the Creative Commons Attribution (CC BY) license (http://creativecommons.org/licenses/by/4.0/).

 *data*

Review

An Interdisciplinary Review of Camera Image Collection and Analysis Techniques, with Considerations for Environmental Conservation Social Science

Coleman L. Little [1], Elizabeth E. Perry [1,]*, Jessica P. Fefer [2], Matthew T. J. Brownlee [1] and Ryan L. Sharp [2]

1   Department of Parks, Recreation, and Tourism Management, Clemson University, 263 Lehotsky Hall, Clemson, SC 29634, USA; colemal@clemson.edu (C.L.L.); mbrownl@clemson.edu (M.T.J.B.)
2   Horticulture and Natural Resources Department, Kansas State University, 2021 Throckmorton, Manhattan, KS 66506, USA; jfefer@ksu.edu (J.P.F.); ryansharp@ksu.edu (R.L.S.)
*   Correspondence: eeperry@clemson.edu

Received: 13 May 2020; Accepted: 3 June 2020; Published: 6 June 2020

Abstract: Camera-based data collection and image analysis are integral methods in many research disciplines. However, few studies are specifically dedicated to trends in these methods or opportunities for interdisciplinary learning. In this systematic literature review, we analyze published sources ($n = 391$) to synthesize camera use patterns and image collection and analysis techniques across research disciplines. We frame this inquiry with interdisciplinary learning theory to identify cross-disciplinary approaches and guiding principles. Within this, we explicitly focus on trends within and applicability to environmental conservation social science (ECSS). We suggest six guiding principles for standardized, collaborative approaches to camera usage and image analysis in research. Our analysis suggests that ECSS may offer inspiration for novel combinations of data collection, standardization tactics, and detailed presentations of findings and limitations. ECSS can correspondingly incorporate more image analysis tactics from other disciplines, especially in regard to automated image coding of pertinent attributes.

Keywords: automated image coding; data collection methods; interdisciplinary learning theory; research methods; systematic literature review; visitor use management

---

1. Introduction

Camera usage is a valuable research tool, particularly due to the breadth of data collection and analysis facilitated by camera technology and related software [1]. In the discipline of environmental conservation social science (ECSS), cameras and associated image data are frequent methods in collecting information on human interactions with the environment [2,3]. Cameras are well suited to examine ECSS concepts and contexts, as image data and associated analyses can be wide ranging and capture similarly broad information. However, camera usage often requires careful attention to detail, a substantial timeframe, and significant researcher involvement, indicating opportunity for more efficient implementation.

Inspiration for more efficient implementation may come from any of the many disciplines that use cameras, yet camera usage and image analysis as a general method has yet to be systematically explored for cross-disciplinary insight and advancement. In this regard, the lens on camera methods remains smudgy. Because lessons from within and beyond ECSS could aid ECSS researchers in better employing camera methods, we present a systematic literature review of camera use and image analysis, framed by the theory of interdisciplinary learning, to examine trends and extract guiding principles for ECSS researchers.

2. Interdisciplinary Learning

There are substantial research benefits to looking beyond a particular discipline for context, inspiration, and new advancements [4]. Examining cross-disciplinary approaches can advance discipline-specific methods by identifying both singular methods and combinations of them applicable to new contexts.

Interdisciplinary learning provides a framework for understanding how and why cross-disciplinary knowledge can benefit a particular discipline [5,6]. The theory of interdisciplinary learning states that combining similar aspects of differing disciplines to reflect ideas and approaches both known and novel to a context is beneficial and effective for promoting rigorous intradisciplinary advancements [6].

Many studies have examined the benefit of looking beyond a particular discipline for *context* inspiration and new advancements [7–9]. Fewer have examined *methods* transferability across disciplines, though ones that have done so have been transformative. One example is the work of Alden, Laxton, Patzer, and Howard [10] on incorporating marketing methods into scientific research to better enact scientific policy advancement. Even fewer have examined camera methods in a cross-disciplinary or interdisciplinary manner, suggesting an area for further development. In ECSS in particular, interdisciplinary knowledge about camera methods remain rather underdeveloped outside of the general references to wildlife cameras being adapted and applied in visitor use management studies [11]. Therefore, we focus on synthesizing camera methods (data collection and resulting image analysis techniques) beyond wildlife and fisheries studies across disciplines to foster interdisciplinary learning in ECSS.

3. Camera Usage as a Research Method

Many types of cameras are used in research, such as handheld digital, field mounted, infrared, underwater, LAN-based, CCTV security, motion-sensing, airplane-affixed, and satellite-based cameras [12]. Analysis methods are correspondingly diverse, including manual coding, digital coding, automated coding, feature detection, and time-lapse sequencing, depending on the research aim [1]. There has been an increasing reliance on camera use as a research method in disciplines including natural, social, and technology sciences [1]. Two themes of camera usage are prominent across the literature: methodological similarities and differences across disciplines and time periods.

3.1. Methods Are Discipline Specific and Discipline Transcending

Camera-related research has both discipline-specific and discipline-transcending methodologies. Specifically, while certain methods are considered reliable practices solely in a particular discipline, others are considered reliable practices (with context-specific modifications) across several disciplines. For example, marine geological research uses boat-mounted cameras to map seafloor features, but other disciplines rarely report using these cameras [13]. However, the remote-sensing camera method of LiDAR is a major component of many environmental subdisciplines, such as agriculture, land use, and climate change research [14].

Discipline-transcending camera methods are typically those that have a longer history of use (an indicator of their reliability), are able to function alongside newer technologies, and are amenable to adaptations for specific contexts and questions [15]. Within ECSS, camera methods are both discipline specific and discipline transcending [15,16]. For example, participant-worn cameras to examine park-based recreation are unique to ECSS but camouflaging field cameras to examine park use has been adapted from other disciplines [15,17,18]. Indeed, many applications of camera methods in ECSS were originally adapted from studies centered on studying wildlife and other non-human animals [19].

3.2. Methods Have Evolved in Diversity and Complexity

Cameras have evolved from centering on large equipment with film and hardcopy photographs into small devices capable of digital images accessible by many computer-based interfaces [20]. As with other technological advancements, the shift in cameras from manual to automated processes and related capability to digitally capture, edit/enhance, and analyze images has increased the utility of cameras to research [20,21]. Manual coding involves someone examining the image data and assigning codes to attributes of interest, whereas automated coding uses analysis software and artificial intelligence to code these attributes [22]. YOLO: Real-time Object Detection [23], WUEPIX [24], and Amazon Rekognition [25] are a few examples of automated image analysis software.

Advancements in technology have had a noticeable impact on how cameras are used in research [20,26]. Early camera usage in research focused on providing visuals to complement evidence described in text format and not necessarily derived from the visual itself [27]. In recent decades, camera usage has shifted to become a method itself [26]. Pre-1995, camera methods focused on film [28] and manual coding [29]. Post-1995, the emphasis has shifted to digital images and automated coding [21], as well as a proliferation of the types of cameras used (e.g., satellite, surveillance). Recent advancements in computer technologies, such as cellular and satellite technologies, and automated image analysis software have further extended the utility of cameras from a research method in small case studies to a tool for big data investigations [30].

4. Research Questions

Despite the numerous research publications showcasing the diversity and complexity of camera methods, and the method's future applicability, there has not been a synthesis of this breadth and its evolution to document patterns of novelty and commonality [31] to facilitate interdisciplinary learning broadly or in ECSS in particular. It appears that inquiry into the subject has focused on a subset of the broad methodology, such as reviewing techniques within facial recognition [32] or remote sensing [33], analyses based on neural network segmentations [34] or classification systems [35], or medical database retrieval accuracy for image data [36]. A review across techniques, analyses, and disciplines appears to be lacking. We address this general need for camera method interdisciplinary learning and devote particular attention to ECSS by focusing on four primary questions:

1. In what contexts have cameras been used in general?
2. In what contexts have cameras been used in ECSS?
3. What are common image collection techniques for image data?
4. What are common analysis techniques for image data?

In synthesizing general and ECSS-specific patterns, we aim to draw conclusions for interdisciplinary learning and related recommendations [37,38].

5. Materials and Methods

We performed a systematic literature review [39], examining studies that used camera methods and image analysis. Our review conformed to PRISMA guidelines [40], modified slightly for study-specific aims (Figure 1). PRISMA guidelines list standard and transparent steps in the harvesting, analyzing, and reporting of data. We followed all steps for the harvesting of data (Figure 1) and reporting on all section/topics in the PRISMA checklist [40]. Modifications to the PRISMA process were in the analyzing of these specific data, as we qualitatively coded a variety of sources and some features of PRISMA's primarily quantitative evaluations of randomized trials did not directly apply to this particular context or framing (e.g., source bias, meta-regressions). This methodology yielded thousands of documents that were systematically sifted to create a subset of documents relevant to our research questions.

The author team defined keyword criteria for inclusion. After gaining general content familiarity through searches for publications, we refined the inquiry to four primary search terms using Boolean operators: "camera*" AND "image*" AND "image analysis*" AND "image data*". To filter the general results from this first broad search, we conducted a series of 15 additional searches, each with an added keyword phrase to these four primary search terms, to focus the inquiry. The additional keyword phrases (e.g., common image analysis software platforms) and key terms related to ECSS used in conjunction with these four primary search terms were "Amazon Rekognition*", "activit*", "artificial intelligence*", "attribute*", "cod*" [for coding-related terms], "distribution*", "Google Vision", "park", "protected area*", "recreation*", "timelapse*", "use level*", "visitor*", "Wuepix*", and "YOLO*".

After an initial query into the utility of multiple databases, three were selected: Agricola, Google Scholar, and Web of Science (Figure 1). These databases were purposefully selected to capture a breadth of sources from peer-reviewed journals to theses/dissertations to management reports. We then conducted the final literature search from November 2018 to January 2019.

Exclusion criteria were used to capture the breadth of sources and disciplines but retain parameters. An overarching exclusion criterion was wildlife and fisheries discipline sources, as the high volume of sources pertaining to camera traps in that discipline would have otherwise overshadowed the sources pertaining to camera and image data in other disciplines. Furthermore, because cameras are a well-established methodology in wildlife and fisheries, reviews of these techniques have already been published [38,41–44]. Therefore, so as to not take away from the ECSS focus of this literature review, 102 sources screened but relating to wildlife and fisheries research were excluded from the final dataset. Relatedly, we did not include "camera trap" or other wildlife-specific terminology in our search terms. Beyond this general criterion, three additional exclusion criteria filtered the results from the remaining relevant sources. Sources must be: (1) published in peer-reviewed journals, as theses/dissertations, as conference proceedings, or as technical reports; (2) written in English; and (3) available to the researchers via full-text online or through Interlibrary Loan.

The assessment of relevance detailed in Figure 1 refined the thousands of sources for study inclusion to 391 (see Supplementary Materials). First, the title and abstract (or similar information if an abstract was not provided) of 3318 keyword search results were examined for initial relevance (i.e., does the title/abstract actually discuss issues germane to the keyword search?). A subset of the author team methodically assessed which specific search terms and related phrasings best fit the scope of the sources, determined the categorization of these sources, and employed consistent practices to systematically assess relevance. Three criteria characterized this process for potential inclusion at this stage: 1. each source had to mention both camera use in research and a corresponding image analysis in its title and/or abstract; 2. each source had to describe research from an image dataset (i.e., no reviews or syntheses); and 3. each source had to consist of more than just a title and abstract (i.e., an actual source had to accompany the title/abstract). The majority of the sources returned via the keyword searches did not contain all three of these characteristics (e.g., camera usage was merely a subsection of a certain procedure outlined rather than a detailed explanation regarding the collection and processing of camera data) and thus were excluded.

This first phase, plus removing duplicates, reduced the relevant sources to 655 for potential inclusion. In the second phase, these sources were downloaded and read in full. The author team divided reading these sources, assessing their relevance, and, if relevant, entering them into the study database. Intercoder reliability measures were employed to minimize discrepancies among data entries [45], with two members of the author team acting as the primary and secondary data enterers, respectively, and performing checks on the others' work. This approach helped increase standardization and decrease individual bias, ensuring that each coder was following a substantially similar approach to entering sources into the database and eliminating non-relevant ones. It did not fundamentally alter the number of sources inputted, but rather the consistent quality of metadata entered about each source. Upon full review, 391 sources (62%) were deemed relevant and entered in the study database. This database captured source metadata (e.g., citation information), camera(s) details used in the research, image analysis technique(s), key study findings related to the topic, and key study findings related to camera use and image analysis. The 264 sources omitted as irrelevant were excluded mainly because they only made tangential reference to cameras and their application, rather than as a method for the study itself.

Following the team's entry of the 391 relevant references into the database, the resulting dataset was coded and analyzed. This analysis was led by the primary and secondary data enterers, as they were most familiar with the data corpus, with assistance from the full team. Six attributes of database entries were qualitatively coded into key themes within each attribute [45]: research discipline, country and continent of study, camera type, camera placement, data collection method, and data analysis method. Other database categories (e.g., publication year, number of image attributes examined) lent themselves to purely quantitative analysis. Descriptive statistics were generated and comprise most of our analysis.

The Supplementary Materials accompanying this manuscript lists the 391 sources analyzed in this systematic literature review, including their citation information and permanent access links (e.g., DOI). Each source has a unique ID: S (for "source") 001–391. Hereon, we reference examples of sources by their unique IDs. This format highlights examples across the breadth of this large dataset while constraining superfluous in-text citations. We encourage readers to examine the supplementary file for citation information for a particular example or across the entire corpus of sources.

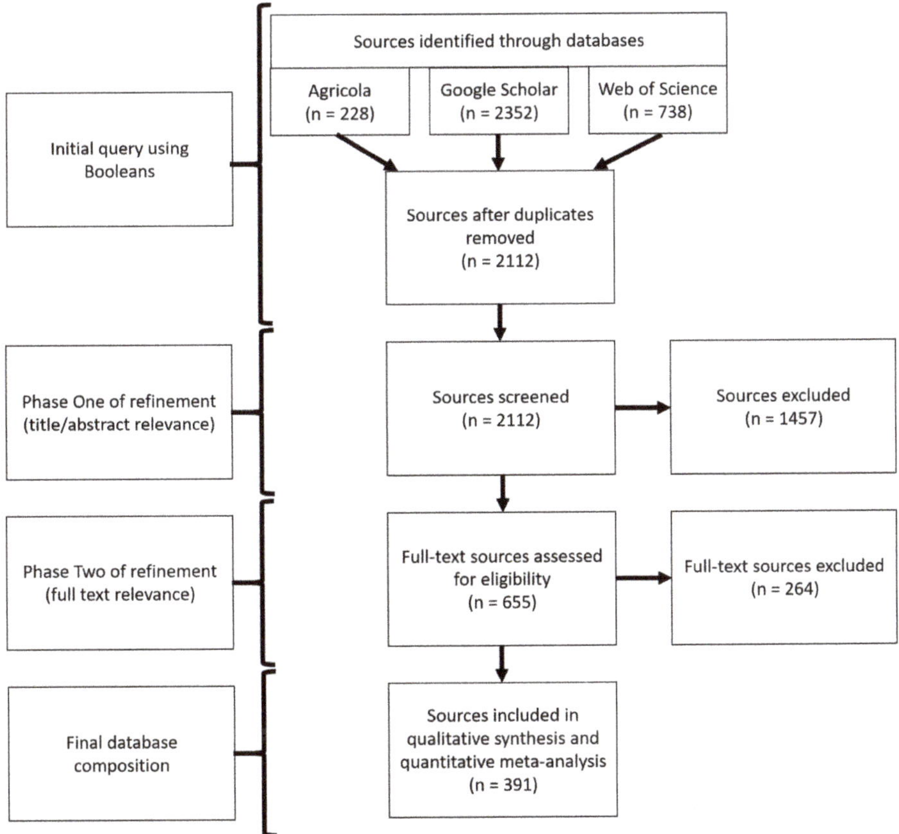

Figure 1. Steps followed to refine the corpus of sources included in this systematic literature review, from initial query to final database. Following this process, citation metadata and six attributes were thematically coded for each of the 391 included sources: research discipline, country and continent of study, camera type, camera placement, data collection method, and data analysis method.

6. Results

6.1. Contexts of Camera Use in General

Cameras have been used and discussed in a variety of contexts: research disciplines, years, and continents (Table 1). The majority of the sources (74%) were peer-reviewed articles, followed by dissertations and theses (20%), reports (5%), and conference proceedings (1%). Fifteen general research disciplines were apparent, which are used as our main grouping criteria throughout this study (Table 1). The four most prevalent were ECSS (21% of the sources), Engineering and Technology (15%), Agriculture (11%), and Computer Science/Programming (10%). The other 11 each accounted for <6% of the publications (Table 1). Examples of the more prevalent disciplines include an ECSS study that used images from drones in England and Portugal to classify sections of protected areas by main use (e.g., wildlife habitat, ecotourism, law enforcement) (S203) and an Engineering and Technology study also using drones, but to test image quality software and facial recognition technology at varying distances and lighting conditions (S140).

Camera use in research has increased substantially in the past 25 years. Publication distribution over time (Figure 2) depicts this increase, especially in the past 10 years for ECSS, Engineering and Technology, Agriculture, and Computer Science/Programming.

The locations for these studies span countries on six continents and some international collaborations (Table 1). Study locations across research disciplines were most common in North America (37%), particularly in the USA, followed by Europe (22%), Asia (19%), Australia and Oceania (6%), multinational/cross-continental (4%), South America (4%), and Africa (3%).

Most of the sources (77%) focused on a sole attribute (e.g., counts or percentage cover of a particular species or landscape formation, detection/recognition of human faces or a particular person). The remainder focused on two (17%), three (2%), 4–10 (3%), or >10 (1%) attributes. The studies that examined 2–10 attributes focused mainly on presence/absence or percent cover of these attributes (e.g., categories of ecosystem services, frequency of chemical compositions). The five publications that focused on >10 attributes mostly concerned different vegetation or land use classes. Although almost all publications listed the year(s) in which these attributes were collected, only 25% listed specific sampling times. These were mainly those in Botany/Plant Science examining vegetation with seasonal foliage (e.g., S199) and those in ECSS examining peak visitor use times (e.g., S239, S362). The number of attributes considered in camera-based studies is an important measure given the opportunities and challenges associated with analysis strategies. The more attributes to characterize in an image, the more difficult and time-consuming analysis becomes, whether manual or automated.

Table 1. Source metrics by research discipline, source type, year published, and continent of study. Cells are listed as valid percentages (%) of the total sources for each research discipline.

	Agriculture	Biology/ Microbiology	Botany/ Plant Science	Computer Science/ Programming	Engineering and Technology	Environmental Biophysical Sciences	Environmental Conservation Social Science	Food Science	Forestry	Geography	Marine Science	Medicine/ Health Science	Other *	Psychology	Urban Studies	Total
Source Type																
Article	96	100	84	58	68	69	72	100	80	32	40	79	89	100	60	75
Conference Proceedings	0	0	0	0	0	0	4	0	0	0	0	0	0	0	20	1
Dissertation/Thesis	2	0	11	33	24	19	24	0	16	68	20	11	5	0	20	19
Report	2	0	5	10	8	12	0	0	4	0	40	11	5	0	0	5
Year																
1985–1989	0	0	0	0	0	0	0	0	0	0	0	11	0	0	0	0
1990–1994	0	0	0	0	0	0	0	0	0	0	0	0	0	0	0	0
1995–1999	2	0	5	0	2	4	1	0	12	0	0	0	0	0	0	2
2000–2004	5	0	11	3	0	4	5	0	16	11	20	0	5	0	0	5
2005–2009	14	88	21	8	20	27	10	36	24	32	0	32	0	17	0	19
2010–2014	39	0	37	0	3	23	27	14	20	37	0	21	26	33	40	21
2015–2019	41	13	26	90	75	42	57	50	28	21	80	37	68	50	60	53
Continent																
Africa	0	0	0	8	0	4	7	0	0	0	0	0	0	0	0	3
Asia	11	0	26	25	34	12	15	21	32	16	20	16	11	0	20	18
Australia/Oceania	4	0	11	0	8	8	11	7	4	0	20	5	5	0	0	7
Europe	33	38	11	28	22	19	22	7	24	5	0	21	32	17	0	21
North America	18	63	37	38	32	50	32	29	36	74	20	53	26	67	80	37
South America	7	0	0	0	2	0	5	36	0	0	0	0	5	0	0	4
International	7	0	0	3	0	4	7	0	0	5	40	0	5	17	0	4
Not Mentioned	20	0	16	0	2	4	1	0	4	0	0	5	16	0	0	5

* Other are research disciplines with ≤3 sources: Architecture, Astronomy, Chemistry, Climatology, Communications, Construction Science, Criminal Justice, Education, Graphic Design, Marketing, and Textiles

88

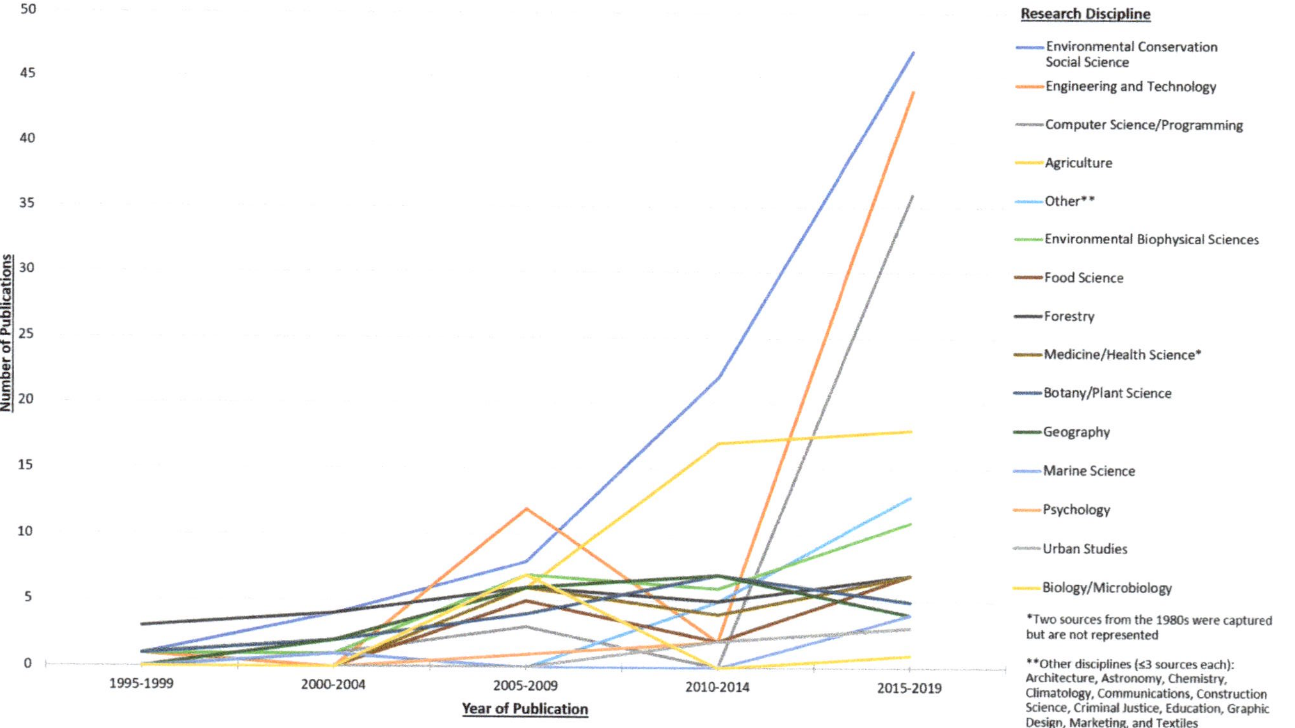

Figure 2. Publication distribution over time (5 year increments from 1995 to 2019) for each research discipline. The research discipline key is presented in the same order as sources, from top to bottom, most to least (i.e., from Environmental Conservation Social Sciences having the highest percentage to Biology/Microbiology having the least).

6.2. Contexts of Camera Use in ECSS

Environmental conservation social science sources were the most numerous by a few different metrics. They were the most frequently represented across articles, conference proceedings, and dissertations/theses (Table 1). The production rate of these publications has been pronounced, especially in the last decade (Figure 2). For example, ECSS publications comprised 31% of the total sources included from 2010–2014 and 23% since 2015. Of the 15 research disciplines represented, ECSS was the only one to have publications concerning all six continents (Antarctica had no studies), as well as international/multinational domains. It was also the most numerous across each continent and context, except in South America where Food Science had one more publication. Almost a quarter (23%) of all the ECSS publications focused on studies in the USA.

Categorical codes were applied to attributes within ECSS studies, to examine the major areas within ECSS that are using camera image analysis. Ten categories emerged: park visitor use management (24%), human–wildlife interactions (22%), recreation ecology (17%), general tourism (9%), public participatory GIS-PPGIS (6%), recreational behavior (6%), sports tourism (including extractive sports, e.g., hunting, fishing) (6%), urban tourism (6%), climate change (2%), and environmental education (1%). Because ECSS is an inherently applied science, all of the categories also encompass a "planning" aspect for managerial use (e.g., park managers, urban planners).

6.3. Common Data Collection Techniques

Almost all, 97% ($n = 380$), of the sources stated at least a general camera type (e.g., webcam, two thermal cameras) and 49% of these detailed the specific camera make and model. For ECSS publications in particular, 41% specified a camera make and model.

Words used to describe the quality of the images obtained in each study were indifferent to positive (e.g., fair, average, decent, good, great, precise, high resolution), with 11% ($n = 42$) of those with a description of image quality forgoing an adjective in favor of listing the pixel resolution. ECSS publications were more apt to describe variability in the images. Whereas this was mostly absent from descriptions in other disciplines, 16% of the ECSS sources with a description noted fuzziness, shakiness, weather-related clarity issues, or, in the case of participatory research, variability according to the user (e.g., S022, S148, S367).

Data were collected through a variety of camera placement techniques. Most of the publications, 92% ($n = 361$), mentioned the primary camera placement technique in their methods: mounted to outdoor fixed location (32%), indoor lab equipment (20%), aircraft/drones (18%), computer (9%), satellite (6%), participatory (participant used in-person or online) (5%), wearable (researcher worn) (5%), watercraft (2%), or vehicle (2%). Of the 80 ECSS publications listing the primary camera placement, 60% were in outdoor fixed locations and 19% used aerial imagery. The aerial imagery for ECSS was mainly obtained through drones (e.g., S137, S160), whereas aerial imagery across the whole dataset was mainly obtained from aircraft-affixed cameras (e.g., S019, S065, S096, S255). ECSS also had the most sources using participant-worn cameras (e.g., S082, S120, S345, S348).

While all placement techniques have generally increased over the past 25 years (Figure 3), increases over the past 10 years are especially pronounced for outdoor fixed, aircraft/drones, and computer mounted techniques. In some cases, data from multiple scales and placements were used. For example, aerial or satellite imagery was paired with ground-truthed transect line images to examine: leafy spurge in wildland areas (S040), proportions of live versus burned or cut vegetation across the western USA (S146), and sources of impact (including recreation) to coral reefs in a marine protected area (S253). Although many camera placement technique usage rates still occupy a relatively small proportion, the general trend is that placement technique diversity is growing, with multiple data collection formats represented. ECSS sources illustrate this trend (Figure 4), with diversity increasing over the past decade even without indoor lab equipment or vehicle placement techniques represented.

The majority (78%; $n = 304$) of sources contained at least one recommendation related to camera-based data collection. Across disciplines, the most common recommendations concerned best practices for using digital cameras when researchers were using fixed/mounted cameras (46%), with a specific recommendation to standardize distance between the camera and object/phenomenon of interest being paramount. Beyond this, specific camera features were noted. For example, an Engineering and Technology paper on combustion behaviors in a coal furnace found that quality high-speed camera features were crucial (S185). The second and third most common recommendations also concerned digital cameras, but specifically those in fixed locations that took automated images outdoors publicly (12%) and covertly (9%), respectfully. Recommendations for publicly located fixed cameras were present in 11 disciplines, indicating interdisciplinary salience, whereas recommendations for covertly located fixed cameras were only present in six disciplines and were especially concentrated (61%) in ECSS. Common examples of recommendations for publicly located cameras included having capacity for nighttime and infrared image capture (e.g., S226, S307), considering the stability of the mount's substrate (e.g., S237, S271), and embedding metadata including GPS location into each image captured (e.g., S018).

An observed pattern in key recommendations by discipline is that some disciplines are highly specialized in a subset of particular camera data collection methods whereas others are more dispersed. We coded 46 different types of camera data collection recommendations. ECSS and Engineering and Technology addressed at least half of these types. At the other end, Biology/Microbiology, Geography, Psychology, Marine Science, and Urban Studies had sources addressing <20% of these types. We collapsed these 46 types into six overarching categories: fixed/mounted (14 methods; 211 sources), held/worn (7 methods; 78 sources), alternate/modified image capture (8 methods; 42 sources), moving (9 methods; 91 sources), multiple (3 methods; 8 sources), and security/surveillance (5 methods; 37 sources) (Table 2). As the distribution in each category suggests, some data collection methods (e.g., multiple cameras) have many recommendations centralized on a few techniques and others (e.g., fixed/mounted cameras) have more dispersed recommendations across many techniques.

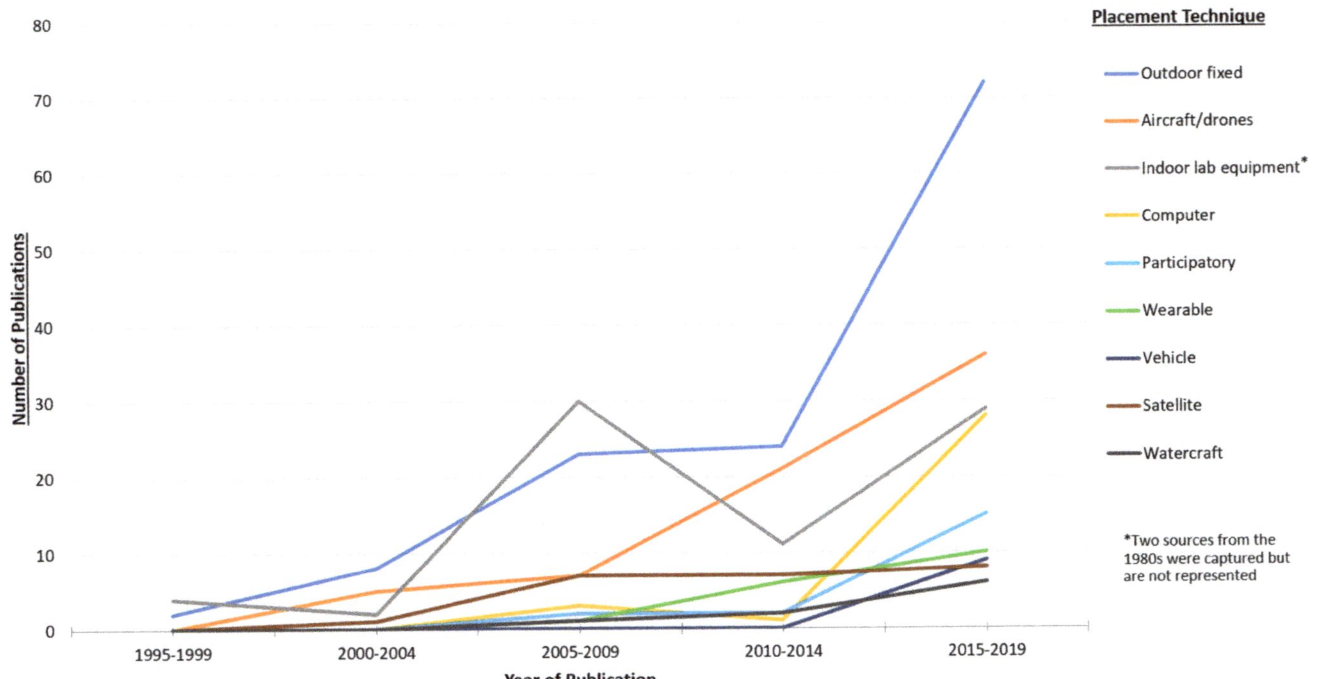

Figure 3. Publication distribution over time (5 year increments from 1995 to 2019) for each camera placement technique. The placement technique key is presented in the same order as sources, from top to bottom most to least (i.e., from outdoor fixed having the highest number to Watercraft having the least).

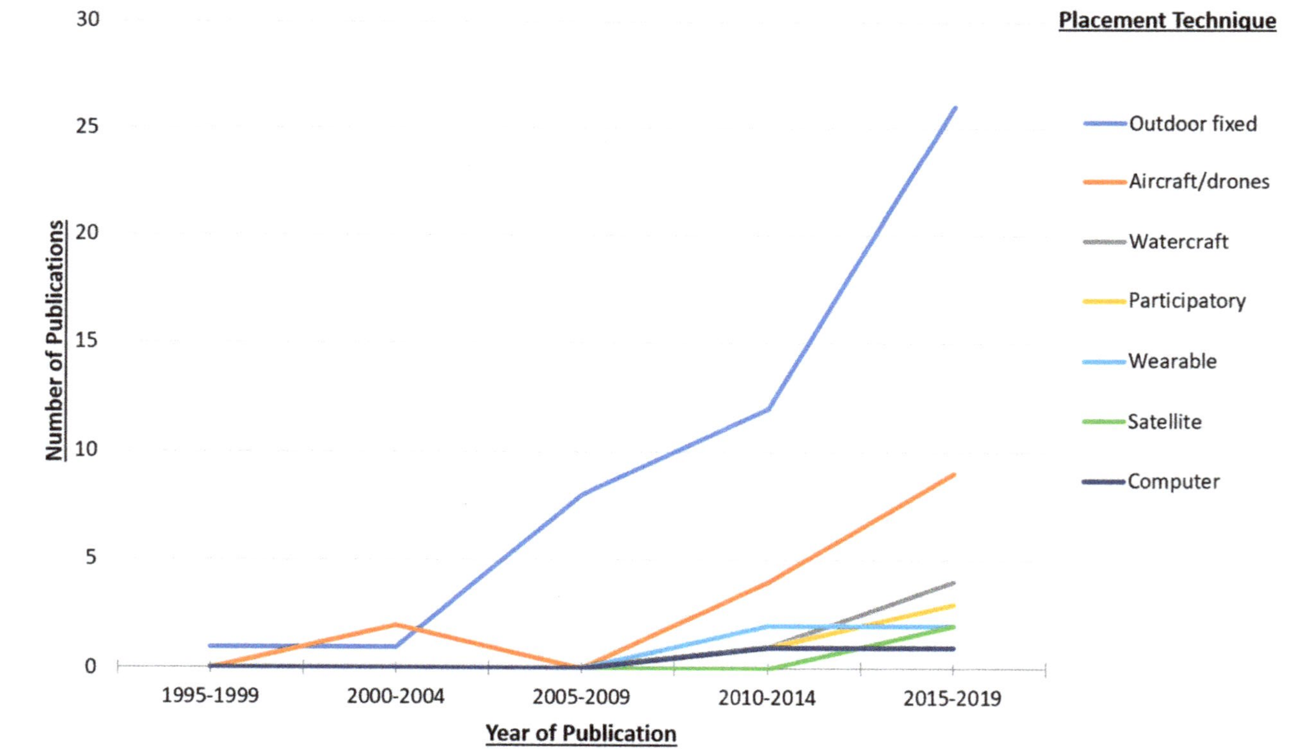

Figure 4. Environmental conservation social science publication distribution over time (5 year increments from 1995 to 2019) for each camera placement technique. The placement technique key is presented in the same order as sources, from top to bottom most to least (i.e., from outdoor fixed having the highest number to Computer having the least).

Table 2. Source metrics by camera placement method and data collection and analysis recommendations. Cells are listed as valid percentages (%) of the total sources for each research discipline.

	Agriculture	Biology/ Microbiology	Botany/ Plant Science	Computer Science/ Programming	Engineering and Technology	Environmental Biophysical Sciences	Environmental Conservation Social Science	Food Science	Forestry	Geography	Marine Science	Medicine/ Health Science	Other *	Psychology	Urban Studies	Total
Camera Placement																
Aircraft	38	13	27	5	16	21	19	14	29	38	0	0	12	0	0	18
Computer	3	0	7	25	23	8	3	7	0	0	0	6	6	0	0	8
Indoor lab equipment	16	75	33	13	30	4	0	43	14	0	0	78	35	33	0	20
Outdoor fixed	38	13	33	35	14	29	60	29	48	6	50	0	12	17	0	34
Participatory	0	0	0	15	2	8	5	0	5	0	0	0	18	17	25	5
Satellite	0	0	0	0	2	21	3	0	5	56	25	0	6	0	75	6
Vehicle	3	0	0	5	7	4	0	0	0	0	0	6	0	0	0	2
Watercraft	0	0	0	0	0	4	6	0	0	0	25	0	0	0	0	2
Wearable	3	0	0	3	7	0	5	7	0	0	0	11	12	33	0	4
Camera Data Collection Recommendations																
Alternate/Modified	6	21	19	2	14	3	5	29	11	8	0	8	10	0	0	9
Fixed/Mounted	45	57	48	59	36	43	50	43	36	23	67	50	29	60	0	46
Held/Worn	6	14	5	17	26	13	15	19	18	0	0	19	33	20	0	17
Moving	27	0	14	7	17	33	22	10	29	69	33	8	14	0	100	19
Multiple	3	0	5	2	0	0	2	0	4	0	0	4	5	0	0	2
Security/Surveillance	12	7	10	14	7	7	6	0	4	0	0	12	10	20	0	7
Camera Data Analysis Recommendations																
Automated	22	0	0	55	55	0	10	29	17	57	33	25	29	50	0	32
Geospatial	22	0	33	5	9	50	27	0	33	43	67	0	0	0	71	20
LiDAR	11	0	0	5	0	0	3	0	0	0	0	0	0	0	0	3
Manual	22	0	33	10	18	50	50	57	33	0	0	75	29	50	14	29
Mixed methods	22	0	33	25	18	0	10	14	17	0	0	0	43	0	14	16

* Other are research disciplines with ≤3 sources: Architecture, Astronomy, Chemistry, Climatology, Communications, Construction Science, Criminal Justice, Education, Graphic Design, Marketing, and Textiles.

6.4. Common Image Analysis Techniques

Only 142 sources (36%) offered data analysis recommendations (Table 2). We coded these recommendations into 44 different analysis procedures, grouped within five more general categories: automated (23 techniques; 46 sources), geospatial (2 techniques; 29 sources), LiDAR (1 technique; 3 sources), manual (12 techniques; 41 sources), and mixed-methods (6 techniques; 23 sources) analyses. Automated techniques included analyses with customizable software such as YOLO, Google Vision, Amazon Rekognition, and eCognition. An example of this is combining a new method of active learning in YOLO with an incremental learning scheme to accurately code vehicle-mounted video camera images (S185). Geospatial techniques focused on particular spatial data attributes, such as types and resolutions of satellite imagery that adequately captured forested, urban, and benthic features (e.g., S014, S035, S129). LiDAR highlights the utility of remote sensing in monitoring long-term impacts of natural processes like the time-lapsed mapping of vegetation growth in forest habitats using LiDAR surveying methods (S280). Manual analysis was concentrated in the labor-intensive process of human coding of primary and secondary (e.g., social media images) data. Although labor-intensive, many sources cited the increased accuracy of the manual coding as preferable over current, accessible automated techniques (e.g., S130, S351) and some offered novel ways for coding large datasets, such as utilizing citizen scientists (e.g., S335). Finally, mixed-methods analyses combined automated and manual techniques, a "human-in-the-loop" approach, to validate automated methods with a sample of human-coded images from the same dataset. A common example used human-in-the-loop approaches to test whether facial recognition software could accurately recognize people, human features, and/or emotions (e.g., S024, S113, S162, S275, S349). As the distribution of techniques and sources across categories implies, some analysis techniques (e.g., geospatial) have many recommendations centralized on a few procedures and others (e.g., automated) are more dispersed across procedures.

The majority of disciplines exhibited concentration of analyses within particular methods. Ten disciplines had at least half of their sources within one category of analysis. Medicine/Health Science was most concentrated, with 75% of its recommendations concerning manual analysis. Many disciplines were concentrated within two analysis categories: Environmental Biophysical Sciences (50% geospatial, 50% manual), Geography (57% automated, 43% geospatial), Marine Science (33% automated, 67% geospatial), Medicine/Health Science (25% automated, 75% manual), and Psychology (50% automated, 50% manual). Agriculture, Computer Science/Programming, and ECSS had all five analysis categories represented. In ECSS, half of the sources had manual coding recommendations (relatively high for the dataset) and only 10% had automated coding recommendations (relatively low for the dataset).

7. Discussion

Our systematic review indicates an increase in the use of camera methods over the past 20 years, and a related proliferation in types of image analyses. However, camera data collection and image analysis techniques have largely developed within disciplines, limiting the ability for collaboration and interdisciplinary learning. Framed by interdisciplinary learning theory, the following synthesizes patterns in camera usage and image analysis, as well as overall best practices and ECSS-specific recommendations.

Although discipline and study-specific contexts will require adaptations, standardized data collection methods and automated analyses can assist in interdisciplinary learning. Technological advancements have facilitated increased camera use and complexity of analyses. Manual coding is more time consuming but requires less sophisticated knowledge of complex software and computer scripts. Several disciplines are utilizing automated analyses and researchers in these disciplines could provide cross-disciplinary guidance for further usage of these analyses. As ECSS uses camera-based data collection but rarely uses automated analysis methods, this discipline in particular could benefit from interdisciplinary collaborations on types of automation and relative benefits and costs.

8. Camera Usage

Few sources make recommendations about camera usage. Those that do tend to focus on standardization techniques for manually taken images. Beyond this specific type of recommendation, our review suggests three areas for best practices: (1) harness the capability of digital datasets to examine multiple locations and attributes, which may be across disciplines; (2) be intentional and specific about documenting study and camera details for other researchers; and (3) examine camera research outside of your own discipline for inspiration.

Although the purposes for image use and study contexts vary across and within disciplines, studies tended to focus on a single attribute obtained from outdoor mounted cameras and in locations concentrated in Europe and North America. Within ECSS, studies most commonly focused on park visitor use management, human–wildlife interactions, and recreation ecology. These patterns suggest an opportunity to expand in geographic settings and to harness automated analysis methods to code beyond a sole attribute. LiDAR and satellite-based camera technology have gained prominence and may offer a means to collect data from more locations without the associated researcher costs of geographic expansions. These techniques also suggest opportunity for researchers in different disciplines to share a common dataset while focusing on attributes of discipline-specific interest. For example, satellite-based image data covering a designated cultural landscape could provide information pertinent to Agriculture and ECSS.

Camera usage should be detailed further to enhance replicability. This could be through additions as simple as stating the specific camera model used and specific data collection timeframe. Metadata could detail image quality beyond simple adjectives, so that other researchers could assess method utility to their contexts. Few papers detailed specific image quality aspects, indicating that a baseline for comparison across camera types might be warranted for standardization (e.g., defined scales and notations).

Some disciplines are more specialist, and some are more generalist. This provides an opportunity to examine novel designs. For example, although ECSS uses the largest diversity of camera placement methods, these tend to be concentrated in fixed and mounted designs. Other disciplines may offer inspiration for using other combinations of methods and placements. Differential LiDAR use across disciplines provides a specific instance of interdisciplinary learning for ECSS. LiDAR is mostly applied in large landscape contexts to classify vegetation growth for natural resources and agricultural studies. Although ECSS has the fastest growth rate of camera method use compared to other disciplines (Figure 2), it has yet to incorporate LiDAR. To date, ECSS largely uses cameras for counting attributes within an image (e.g., visitors, vehicles, boats) to understand types and frequencies of human behaviors in an environment. ECSS also uses cameras to understand place-based experiences through participatory camera methods. Both applications tend to occur on the site, rather than landscape, scale. Sub-disciplines within ECSS, such as recreation ecology, might benefit from using LiDAR to detect landscape level differences in ground cover over recreational uses and longer temporal scales.

9. Image Analysis

Sources used a range of image analysis techniques within automated, geospatial, manual, LiDAR, and mixed methods but only approximately one-third (35%) offered recommendations for image analysis. We offer three fundamental practices for researchers to enhance interdisciplinary learning opportunities: (1) list and provide critical analysis of image analysis methods; (2) examine image analysis techniques beyond those typically utilized in a particular discipline; and (3) standardize guidelines for certain analysis techniques, particularly ones that are discipline specific but may have applicability across disciplines.

Disciplines favor particular categories of image analysis. This concentration implies disciplinary expertise but also areas for more creative interdisciplinary insight. Several disciplines continue to rely on manual coding techniques (e.g., ECSS, Medicine/Healthcare), while others have developed automated processes (e.g., Agriculture). This discrepancy reflects a lack of interdisciplinary sharing and but also a necessary emphasis on case study approaches. For example, many ECSS studies that use outdoor fixed cameras to estimate visitor use would benefit from automated analyses of image attributes across these large datasets, while other ECSS studies that use participant-worn cameras to gain in-depth visitor experience information would be better off manually coding their images. Although these differences depend on the study purpose and approach, software to facilitate automated coding and guidelines for manual coding of image data are both needed.

Just as multiple disciplines have benefited from guidelines for qualitative data coding and statistical analysis software use, guidelines for both manual and automated image coding would provide interdisciplinary standards and efficiencies. ECSS is still primarily dependent on manual coding. Although there have been attempts within ECSS to codify guidelines for manual image coding [46,47], these sources have yet to be cited regularly within ECSS or at all in other disciplines. Examining methods of automated image analysis and forming partnerships with those who have employed such methods or understand the technology behind them could open up further relevant inquiries on ECSS image datasets. The diversity of automated analysis techniques captured in this study suggests another area for interdisciplinary collaboration, guidelines development, and standards definition, so that researchers can more easily recognize which techniques are best suited for study purposes. This again underscores the importance of interdisciplinary learning, where examining multiple means of image analysis may lend creative insight into how one discipline could learn techniques from another.

10. Limitations

Keyword searches were crafted by this team of ECSS researchers and criteria for source inclusion in this review may reflect biases that would not be apparent if conducted by other researchers. However, we took steps to minimize subjectivity such as using an established method for systematic literature reviews and validating reliability among the research team. Discipline-specific camera usage and analysis jargon and knowledge may have been inadvertently omitted on account of the ECSS researchers' unfamiliarity with these technical terms and thus led to an underrepresentation of particular areas in our findings. Again, we have attempted to lessen this concern through a standardized keyword search using basic, non-technical terms that transcend disciplines and by examining sources for multiple points of relevance.

11. Future Research

The findings of this review highlight four pertinent avenues of future research in general and within ECSS. First, a streamlined method for calculating and reporting the distance between a camera and the attributes of interest would be an interdisciplinary contribution to standardization. Second, ECSS researchers using cameras in studies could test the applicability of LiDAR to questions and contexts within the ECSS discipline. Third, a review of analysis strategies for images posted on online platforms (e.g., social media) could also be conducted and more reliable analysis strategies, particularly the development of a program that would reduce the burden of manual analysis and allow for more images to be included in the analysis, could be developed. Thus far, studies centered on social media images mostly involved geo-tagged images or captions rather than actual image content. Fourth, participant-generated image data should be examined independently, as this data collection technique is uniquely and intentionally less under researcher control.

12. Conclusions

This study assessed a large dataset of sources for enhanced methods pertaining to camera usage and image analysis in general and in ECSS in particular. Using a systematic literature review and interdisciplinary learning theory, this study identified areas of disparity and areas for enhanced collaboration. Six best practices for camera usage and image analysis emerged: examining multiple attributes/phenomena, being intentionally specific in documenting camera details and placement, sourcing methods beyond a specific discipline for novel approaches, critiquing image analysis methods used, examining possibilities for interdisciplinary analysis techniques, and standardizing analysis methods at least within disciplines. The ECSS focus of the study revealed that the discipline is well positioned to be a center of standardization in some regards (e.g., manual coding guidelines) but could benefit from interdisciplinary collaborations (e.g., use of LiDAR). This review provides a snapshot of the wide lens of camera-based methods in research and underscores the need for assessing the diversity of this method, especially as it continues to diversify and proliferate across disciplines and contexts.

Supplementary Materials: The following are available online at http://www.mdpi.com/2306-5729/5/2/51/s1, Data: Corpus of Sources with Citation Metadata.

Author Contributions: C.L.L. conducted the initial inquiry and phase one of the review, was task manager of all components of phase two (including leading source input into the database), assisted in data analysis, and drafted major portions of the manuscript. E.E.P. was project manager of all components, co-led inputting sources, led the data analysis, drafted major portions of the manuscript, and edited the final manuscript. J.P.F. inputted sources, drafted major portions of the manuscript, and edited the final manuscript. M.T.J.B. conceptualized the project, inputted sources, drafted minor portions of the manuscript, and provided guidance on framing the manuscript. R.L.S. inputted sources and provided guidance on framing the manuscript. All authors have read and agreed to the published version of the manuscript.

Funding: This research received no external funding.

Conflicts of Interest: The authors declare no conflict of interest.

References

1. Al-Rousan, T.; Masad, E.; Tutumluer, E.; Pan, T. Evaluation of image analysis techniques for quantifying aggregate shape characteristics. *Constr. Build. Mater.* **2007**, *21*, 978–990. [CrossRef]
2. Anderson, K.; Gaston, K.J. Lightweight unmanned aerial vehicles will revolutionize spatial ecology. *Front. Ecol. Environ.* **2013**, *11*, 138–146. [CrossRef]
3. Cox, M. A basic guide for empirical environmental social science. *Ecol. Soc.* **2015**, *20*. [CrossRef]
4. Hazen, D.; Puri, R.; Ramchandran, K. Multi-camera video resolution enhancement by fusion of spatial disparity and temporal motion fields. In Proceedings of the Fourth IEEE International Conference on Computer Vision Systems (ICVS'06), New York, NY, USA, 4–7 January 2006; p. 38. [CrossRef]
5. Mansilla, V.B. Interdisciplinary learning: A cognitive-epistemological foundation. In *The Oxford handbook of Interdisciplinarity*, 2nd ed.; Frodeman, R., Ed.; Oxford University Press: Oxford, UK, 2017. [CrossRef]
6. Spelt, E.J.H.; Biemans, H.J.A.; Tobi, H.; Luning, P.A.; Mulder, M. Teaching and learning in interdisciplinary higher education: A systematic review. *Educ. Psychol. Rev.* **2009**, *21*, 365. [CrossRef]
7. Liu, J.-S.; Huang, T.-K. A project mediation approach to interdisciplinary learning. In Proceedings of the Fifth IEEE International Conference on Advanced Learning Technologies (ICALT'05), Kaohsiung, Taiwan, 5–8 July 2005; pp. 54–58. [CrossRef]
8. Johnson, D.T.; Neal, L.; Vantassel-Baska, B.J. Science curriculum review: Evaluating materials for high-ability learners. *Gift. Child Q.* **1995**, *39*, 36–44. [CrossRef]
9. Haigh, W.; Rehfeld, D. Integration of secondary mathematics and science methods courses: A model. *Sch. Sci. Math.* **1995**, *95*, 240. [CrossRef]
10. Alden, D.S.; Laxton, R.; Patzer, G.; Howard, L. Establishing cross-disciplinary marketing education. *J. Mark. Educ.* **1991**, *13*, 25–30. [CrossRef]
11. Dimitropoulos, G.; Hacker, P. Learning and the law: Improving behavioral regulation from an international and comparative perspective. *J. Law Policy* **2016**, *25*, 473–548.
12. Kucera, K.; Harrison, L.M.; Cappello, M.; Modis, Y. Ancylostoma ceylanicum excretory–secretory protein 2 adopts a netrin-like fold and defines a novel family of nematode proteins. *J. Mol. Biol.* **2011**, *408*, 9–17. [CrossRef]
13. Menzie, C.; Ryther, J.; Boyer, L.; Germano, J.; Rhodes, D. Remote methods of mapping seafloor topography, sediment type, bedforms, and benthic biology. *OCEANS* **1982**, *82*, 1046–1051. [CrossRef]
14. Schuckman, K.; Raber, G.T.; Jensen, J.R.; Schill, S. Creation of digital terrain models using an adaptive Lidar vegetation point removal process. *Photogramm. Eng. Remote Sens.* **2002**, *68*, 1307–1314.

15. An, F.-P. Pedestrian re-recognition algorithm based on optimization deep learning-sequence memory model. *Complexity* **2019**, *2019*, 1. [CrossRef]
16. Su, C.; Zhang, S.; Xing, J.; Gao, W.; Tian, Q. Deep attributes driven multi-camera person re-identification. In *Computer Vision—ECCV 2016*; Lecture Notes in Computer Science; Springer: Cham, Switzerland, 2016; Volume 9906. [CrossRef]
17. Marion, J.L. A review and synthesis of recreation ecology research supporting carrying capacity and visitor use management decisionmaking. *J. For.* **2016**, *114*, 339–351. [CrossRef]
18. Peterson, B.; Brownlee, M.; Sharp, R.; Cribbs, T. Visitor Use and Associated Thresholds at Buffalo National River. In *Fulfillment of Cooperative Agreement No. P16AC00194*; Technical report submitted to the U.S. National Park Service; Clemson University: Clemson, SC, USA, 2018.
19. Schmid Mast, M.; Gatica-Perez, D.; Frauendorfer, D.; Nguyen, L.; Choudhury, T. Social sensing for psychology: Automated interpersonal behavior assessment. *Curr. Dir. Psychol. Sci.* **2015**, *24*, 154–160. [CrossRef]
20. Kharrazi, M.; Sencar, H.T.; Memon, N. Blind source camera identification. In Proceedings of the 2004 International Conference on Image Processing, ICIP '04, Singapore, 24–27 October 2004; Volume 1, pp. 709–712. [CrossRef]
21. Huang, A.S.; Bachrach, A.; Henry, P.; Krainin, M.; Maturana, D.; Fox, D.; Roy, N. Visual odometry and mapping for autonomous flight using an RGB-D Camera. In *Robotics Research*; Christensen, H.I., Khatib, O., Eds.; Springer: Berlin/Heidelberg, Germany, 2017; Volume 100, pp. 235–252. [CrossRef]
22. Bente, G. Facilities for the graphical computer simulation of head and body movements. *Behav. Res. Methods Instrum. Comput.* **1989**, *21*, 455–462. [CrossRef]
23. Alvar, S.R.; Bajić, I.V. MV-YOLO: Motion vector-aided tracking by semantic object detection. *arXiv* **2018**, arXiv:1805.00107.
24. Staab, J. Applying Computer Vision for Monitoring Visitor Numbers—A Geographical Approach. Master's Thesis, University of Wurzburg, Heidelberg, Germany, 2017. Available online: https://www.researchgate.net/publication/320948063_Applying_Computer_Vision_for_Monitoring_Visitor_Numbers_-_A_Geographical_Approach (accessed on 7 June 2020).
25. Chouinard, B.; Scott, K.; Cusack, R. Using automatic face analysis to score infant behaviour from video collected online. *Infant Behav. Dev.* **2019**, *54*, 1–12. [CrossRef]
26. Fraser, C.S. Digital camera self-calibration. *ISPRS J. Photogramm. Remote Sens.* **1997**, *52*, 149–159. [CrossRef]
27. Tatsuno, K. Current trends in digital cameras and camera-phones. *Sci. Technol. Q. Rev.* **2006**, *18*, 35–44.
28. English, F.W. The utility of the camera in qualitative inquiry. *Educ. Res.* **1988**, *17*, 8–15. [CrossRef]
29. Park, J.-I.; Yagi, N.; Enami, K.; Aizawa, K.; Hatori, M. Estimation of camera parameters from image sequence for model-based video coding. *IEEE Trans. Circuits Syst. Video Technol.* **1994**, *4*, 288–296. [CrossRef]
30. Velloso, E.; Bulling, A.; Gellersen, H. AutoBAP: Automatic coding of body action and posture units from wearable sensors. In Proceedings of the 2013 Humaine Association Conference on Affective Computing and Intelligent Interaction, Geneva, Switzerland, 2–5 September 2013; pp. 135–140. [CrossRef]
31. Rust, C. How artistic inquiry can inform interdisciplinary research. *Int. J. Des.* **2007**, *1*, 69–76.
32. Zhao, W.; Chellappa, R.; Phillips, P.J.; Rosenfeld, A. Face recognition: A literature survey. *ACM Comput. Surv.* **2003**, *35*, 399–459. [CrossRef]
33. Blaschke, T. Object based image analysis for remote sensing. *ISPRS J. Photogramm. Remote Sens.* **2010**, *65*, 2–16. [CrossRef]
34. Pal, N.R.; Pal, S.K. A review of image segmentation techniques. *Pattern Recognit.* **1993**, *26*, 1277–1294. [CrossRef]
35. Lu, D.; Weng, Q. A survey of image classification methods and techniques for improving classification performance. *Int. J. Remote Sens.* **2007**, *28*, 823–870. [CrossRef]
36. Muller, H.; Michoux, N.; Bandon, D.; Geissbuhler, A. A review of content-based image retrieval systems in medical applications—Clinical benefits and future directions. *Int. J. Med. Inform.* **2004**, *73*, 1–23. [CrossRef]
37. Kelly, P.; Marshall, S.J.; Badland, H.; Kerr, J.; Oliver, M.; Doherty, A.R.; Foster, C. An ethical framework for automated, wearable cameras in health behavior research. *Am. J. Prev. Med.* **2013**, *44*, 314–319. [CrossRef]
38. Meek, P.D.; Ballard, G.; Claridge, A.; Kays, R.; Moseby, K.; O'Brien, T.; Townsend, S. Recommended guiding principles for reporting on camera trapping research. *Biodivers. Conserv.* **2014**, *23*, 2321–2343. [CrossRef]
39. Pickering, C.M.; Byrne, J. The benefits of publishing systematic quantitative literature reviews for PhD candidates and other early career researchers. *High. Educ. Res. Dev.* **2014**, *33*, 534–548. [CrossRef]

40. Moher, D.; Liberati, A.; Tetzlaff, J.; Altman, D.G.; PRISMA Group. Preferred reporting items for systematic reviews and meta-analyses: The PRISMA Statement. *PLoS Med.* **2009**, *6*, e1000097. [CrossRef] [PubMed]

41. Burton, A.C.; Neilson, E.; Moreira, D.; Ladle, A.; Steenweg, R.; Fisher, J.T.; Boutin, S. Wildlife camera trapping: A review and recommendations for linking surveys to ecological processes. *J. Appl. Ecol.* **2015**, *52*, 675–685. [CrossRef]

42. Rovero, F.; Marshall, A.R. Camera trapping photographic rate as an index of density in forest ungulates. *J. Appl. Ecol.* **2009**, *46*, 1011–1017. [CrossRef]

43. Scotson, L.; Johnston, L.R.; Iannarilli, F.; Wearn, O.R.; Mohd-Azlan, J.; Wong, W.M.; Frechette, J. Best practices and software for the management and sharing of camera trap data for small and large scales studies. *Remote Sens. Ecol. Conserv.* **2017**, *3*, 158–172. [CrossRef]

44. Trolliet, F.; Vermeulen, C.; Huynen, M.C.; Hambuckers, A. Use of camera traps for wildlife studies: A review. *Biotechnologie Agronomie Société et Environnement* **2014**, *18*, 446–454.

45. Saldana, J. *The Coding Manual for Qualitative Researchers*, 2nd ed.; Sage Publishing: Los Angeles, CA, USA, 2013.

46. Balomenou, N.; Garrod, B. Photographs in tourism research: Prejudice, power, performance, and participant-generated images. *Tour. Manag.* **2019**, *70*, 201–217. [CrossRef]

47. Rose, J.; Spencer, C. Immaterial labour in spaces of leisure: Producing biopolitical subjectivities through Facebook. *Leis. Stud.* **2016**, *35*, 809–826. [CrossRef]

© 2020 by the authors. Licensee MDPI, Basel, Switzerland. This article is an open access article distributed under the terms and conditions of the Creative Commons Attribution (CC BY) license (http://creativecommons.org/licenses/by/4.0/).

MDPI
St. Alban-Anlage 66
4052 Basel
Switzerland
Tel. +41 61 683 77 34
Fax +41 61 302 89 18
www.mdpi.com

Data Editorial Office
E-mail: data@mdpi.com
www.mdpi.com/journal/data

www.ingramcontent.com/pod-product-compliance
Lightning Source LLC
LaVergne TN
LVHW070623170726
843515LV00004B/159